Leading With Marketing

The Resource for Creating, Building and Managing Successful Architecture/Engineering/ Construction Marketing Programs

Brian Gallagher and Kimberly Kayler

AuthorHouse™
1663 Liberty Drive
Bloomington, IN 47403
www.authorhouse.com
Phone: 1-800-839-8640

First published by AuthorHouse 11/19/2009

ISBN: 978-1-4490-3967-7 (sc)

Library of Congress Control Number: 2009912524

Printed in the United States of America
Bloomington, Indiana

This book is printed on acid-free paper.

Acknowledgements:

"I'd like to dedicate this book to my wonderful wife, Lauren, and my two incredible kids, Emily and Shane. I also would like to thank to my outstanding mentors, colleagues and friends, particularly Jack Wyatt, Matt Banes and Peter Emmons that I've had the opportunity to work with through the years." – Brian Gallagher

"This book is dedicated to the wonderful mentors I've had over the years in the AEC industry, particularly Anne Ellis, Dianne Bret Harte and Bob Foley. I also want to thank my best friend and husband John, as well as my two sons, Elec and Joel – each of whom makes the stressful days a bit easier with their encouragement and vision for life." –Kimberly Kayler

Author's Note:

Special thanks to Wendy Ward, Clare Martin, Ashley Kizzire, Stephanie Brown and Kristen Miller of Constructive Communication for their insight on many of the topics found in this book. Wendy is the author of much of the section on the basics of advertising, getting the most out of your association membership and creating effective project sheets. Some of her viewpoints can also be found in the section on ROI and brand identity. Ashley's viewpoints are found in the section on how to secure speaking engagements, while Clare's insight can be found on the section related to newsletters. Stephanie added much insight related to web and social media strategies and tactics. Finally, Kristen's insight on graphic design can be found in a few places in the book.

Introduction (Purpose)

Simply put, marketing is the art and science of creating awareness, delivering value, selling your product or service, satisfying needs, and getting them to buy again. Marketing includes a range of activities, including advertising, public relations, direct marketing, measuring return on investment (ROI), communicating during a crisis, and the Internet. While marketing is a critical component of every business, it is one of the least understood and underutilized components, particularly within architecture/engineering/construction (AEC) firms.

An organization's business development and marketing functions must be responsive and flexible to deliver results consistently. Flexibility is only the beginning; business development and marketing also must be accountable for providing a measurable return on investment. Utilizing metrics can help business leaders better understand the effect marketing activities can have on driving profits. Less than robust economic conditions provide an excellent opportunity to validate the effectiveness of strategies and programs. Within successful AEC firms, marketing has a *leadership* role. Marketing's role is to *lead* the organization to choose which markets to focus on, how to target them, how to communicate with them and how to win work. Marketing's strategic role may include creating brand awareness, promoting products and services, differentiating the company, targeting specific project types, identifying sales opportunities, supporting the sales function, and most important, supporting and complementing the strategic mission of the company. Companies are constantly changing strategic focuses, right-sizing and reallocating operating budgets. Marketing isn't immune to adjustments and changes to its programs, projects and budgets, but leading companies maintain a strict focus on marketing.

Measurements of marketing activities represent a growing trend in business. Certain initiatives, such as telemarketing and direct mail, are more easily measured than others. Other programs, such as advertising and public relations, are a little more nebulous and difficult to measure directly. In addition, it can be difficult to measure the value of long-term and existing marketing programs. While this doesn't mean programs should be halted, management should review the activities with a focus on measurement and ROI. To that end, there are various tools and resources available to benchmark marketing activities and projects to ensure the company's targeted ROI is met or exceeded.

The marketing planning and budgeting process is an excellent starting point to review the relationship between marketing activities and sales efforts. The two should be complementary. Setting clear and measurable goals will enhance the overall effectiveness of sales and marketing efforts.

Finally, it is crucial to periodically review the marketing strategy and budget to maintain consistency with reviews of other operating departments. These reviews will confirm that the marketing function is consistent with the company's overall mission. The reviews will also help determine if marketing dollars are being utilized effectively and efficiently and will ensure the company is maximizing its return on investment.

How to Use This Book

"To be a good marketer, you have to know all these tactics, and then you've got to select the right tactics to win your particular battle." -Al Reis

Marketing is an integral part of the business world, however, too many people in the AEC industry do not understand how to use and apply marketing tools and strategies that can help them establish a leadership position and have an impact on their bottom-line. But, today's marketplace is more competitive than ever, and decision-makers are savvier than ever. If there is ever a time to take the lead and embrace a strategic and integrated marketing program, it is now.

This is not another book about how to write a marketing plan. The mission of this book is to help change the way you think about marketing, so your firm can select the right target audiences, communicate with them effectively, and create profitable opportunities. It is designed specifically as a tool for professionals in the AEC fields that are providing leadership to the marketing efforts, with a complete overview of marketing process, tools, systems and best practices. Companies that market services such as architecture, engineering and construction services; as well as companies that sell products, such as materials, systems, and other products used in construction, will find the marketing principles outlined in this book to be beneficial. Based on years of actual experience with a diverse variety of companies and associations in the AEC industry, this book is designed to bridge the gap between traditional marketing concepts and emerging marketing trends, while providing a contextual link to the construction business.

While there are many books that go into great depth on any one of the topics discussed, this book is designed to provide a good solid overview of marketing and should be read by a broad spectrum of your team members. After all, marketing should not be one person or even a department. To be truly successful, the marketing orientation must start at the top of the organization and permeate the entire company.

The following pages detail all of these concepts in detail and provide you with the tools to succeed in leading the marketing function in your organization, and ultimately help you become a leading firm in your market. The information has been assembled in a format to provide you a general background of marketing strategies and techniques, customized to the AEC industry, in an easy-to-use fashion. The book also includes *Jargon Junction*, which provides additional information on terms, topics and acronyms used. We've attempted to bridge the gap between traditional and new marketing as well as plant seeds and stimulate thinking.

About the Authors

Brian Gallagher

Brian Gallagher has served in executive level marketing positions with industry leading engineering, construction and manufacturing organizations. In his roles with O'Neal, Inc, Structural Group and Williams Scotsman, Gallagher has provided strategic leadership to the sales and marketing teams. His extensive background and knowledge of the construction industry enables innovative and effective marketing programs. He frequently writes and speaks on marketing topics. In addition, he has co-founded a marketing and sales consultancy, GBM Marketing and launched several industry-specific web portals.

Gallagher holds a bachelor's degree in Marketing from Towson University and an M.B.A. from Loyola College. He also has served as an Adjunct Assistant Professor of Marketing at Loyola College. He has served in leadership roles for various committees with the American Concrete Institute, the Post-Tensioning Institute, the Concrete Industry Management Program, and other organizations. He also is a member of the Society of Marketing Professional Services.

Contact Information:

Brian Gallagher
864-551-0362
bgallagher@LeadingWithMarketing.com

Kimberly Kayler, CPSM, CSI

President and Founder
Constructive Communication, Inc.

With a journalism degree and a decade of high-level experience serving engineering, architecture and construction firms as a corporate marketing executive, as well as experience working for a full-service advertising/marketing communications agency, Kayler started Constructive Communication, Inc. in 2001 to serve the needs of technical and professional service firms. Clients include five international concrete associations; a variety of general contractors, engineers and architects from around the country; as well as firms in the aerospace, chemical and industrial sectors. Services provided by the growing Constructive Communication, Inc. team include technical writing, proposal development, public and media relations, social media and marketing/communications.

The author of more than 1,250 published articles on a variety of concrete, construction, design, marketing and other technical subjects, Kayler was the first to earn the Certified Professional Services Marketer designation in the state of Ohio from the Society of Marketing Professional Services. She is a frequent speaker on technical marketing and public relations and is a registered provider through the American Institute of Architects. Although she is a graduate of the University of Arizona, she now calls Columbus, Ohio home. She has an M.B.A. from Capella University with a special emphasis in Leadership. She is a member of the Society for Marketing Professional Services, the Construction Specifications Institute and the Women Construction Owners & Executives. Her firm is a Certified Women's Business Enterprise as well as a certified participant in Ohio's Encouraging Diversity, Growth and Equity (EDGE) program through the Equal Opportunity Division of the Ohio Department of Administrative Services. She is a volunteer with the Junior Achievement program and she serves on the Board of Director for the Dublin Foundation, the Dublin Convention & Visitors Bureau, as well as the Small Business Council of the Columbus Chamber of Commerce.

Contact Information:

Kimberly Kayler, CPSM
President
Constructive Communication, Inc.
614-873-6706, voice
kkayler@constructivecommunication.com
Visit www.constructivecommunication.com

Leading With Marketing

How do successful AEC firms become leaders in their markets? Superior customer service, focus on niches, a deep understanding of customer needs, creation of value, impeccable quality, exceptional services, great relationships, a strong brand, a stellar reputation...the list could go on and on.

While there may not be a consistent formula for success for all leading AEC firms, a common characteristic is marketing. These firms don't view marketing as an expense, but as an investment.

Leading With Marketing embodies how a company approaches their business, their marketplace, and their customers. When companies lead with marketing, they choose which markets to target, what services to offer, how to differentiate, how to communicate, and how to win.

As a key leader or marketing professional in your organization, your responsibility is to provide leadership to the marketing process. So how can *Leading with Marketing* help your business?

- Marketing can help frame the vision and direction for your company.
- Marketing must be a critical component of the strategic planning process by helping provide a critical link to the external environment.
- Marketing can help provide direction and lead your company to focus on certain markets by researching, targeting and segmenting markets.
- Marketing can help you understand customer needs.
- Marketing can help define your competitive advantage.
- Marketing can help differentiate your offering and define your value proposition.
- Marketing can define strategies targeting and communicating with your audiences.
- Marketing can implement marketing initiatives and programs.
- Marketing can provide a feedback loop with your customer.

- Marketing can help link marketing plans, sales plans and account plans.
- Marketing can demonstrate a return on investment.

Whether it is handled by a department or an individual, marketing *is* a leadership function in every AEC firm.

Introduction to AEC Marketing

Ultimately, the role of marketing in an AEC firm is a summation of activities designed to identify opportunities, attract clients, win projects, satisfy client needs and win more work. However, marketing in the AEC industry presents many unique challenges. Professionals in the AEC industry may sell products, services, or a combination of the two. Selling products is vastly different from selling services. Each requires distinct processes and skill set. In both cases, marketing can serve as an enabler of the sales process.

The majority of people holding marketing or business development roles in the industry do not have formal marketing training, rather, a technical background. There are many benefits to technical staff members holding marketing/sales roles, specifically their ability to intimately understand what they are marketing and selling. However, without marketing expertise, their efforts may be haphazard or ineffective.

One of the other challenges marketers in this industry face is the lack of training related to marketing and sales in the AEC world. Most business programs, whether undergraduate or advanced, focus on business-to-consumer (B2C) marketing. Or, even if they cover business-to-business (B2B), little time is spent on professional services. Selling/marketing a professional service is vastly different than marketing a product. This is particularly true in the construction industry that is focused on generating revenue based on projects. In addition, business-to-business implies that businesses buy from businesses. In reality, people in business buy from other people in business. Therefore, marketing must be designed to communicate with and influence people.

Yet another challenge for marketing in the AEC industry is the multitude of decision-makers who must be addressed. For each opportunity, architects, engineers and owners need to be addressed, and they need to be reached with different messages. In addition, each organization may have multiple decision makers and influencers. Identifying each person's role in the process can be challenging.

While many AEC firms regularly embark on a formal business planning process, marketing is too often left out of the process.

Oftentimes, when marketing is excluded, firms fail to develop a marketing strategy and a formal marketing plan that outlines objectives, goals and strategies, tactics, and measurement. Marketing is a critical part of a firm's success and should be integrated at all levels of the organization.

Finally, because of the lack of understanding and formal training about marketing, there is often confusion about the difference between marketing and sales. While larger firms may have a sales staff(s) and a marketing professional(s), too often the roles are lumped together. While marketing and sales are both part of the business development function, it is important to understand the difference between the two. ***Marketing's role*** is to help condition the environment for a sale by systematically identifying needs and wants, then develop and implement a plan to communicate potential solutions to satisfying that demand. ***Sales' role*** is to focus on building and maintaining relations that lead to the sale of profitable opportunities. To be truly effective, marketing and sales professional must work very closely together, both as part of a firm's business development process.

Despite these challenges, there is great opportunity for those currently in or seeking to embark on a sales or marketing career in the AEC industry. Even for firms that don't have a dedicated marketing function, an orientation to marketing is essential. Technical professionals need to understand how their roles, actions, and client interactions leave an impression of the firm that has an impact on potential future work. The following guidelines in this book provide a template for success.

Table of Contents

SECTION 1: STRATEGIC MARKETING

Strategic Marketing and the Marketing Plan

We've all heard the axiom "If you fail to plan, you plan to fail." This is certainly true when it comes to marketing efforts. Without a review, synthesis and then documentation of internal and external factors, objectives, strategies and tactics, there is no way to measure your return on investment. The marketing planning process must be an extension of the business planning process.

For AEC firms, marketing must support the business plan and strategic objectives of the company. Marketing strategies are most effective when they are an integral part of the overall organization's strategy. For firms in the construction industry, marketing is about projects. Architects, engineers, and contractors earn revenue by successfully completing projects, and companies sell their products for use on projects. The marketing efforts of an AEC firm must be built around strategies and actions to position the firm to get more projects. The business development process, which includes sales and marketing, should help firms identify the types of projects to target, develop a strategy to capture the projects, and help build a foundation for satisfying clients and getting more of there work.

First, we'll start with a discussion on marketing strategy. In *Kotler on Marketing*[1], Phillip Kotler defined marketing's role as understanding, creating, communicating and delivering value. Thus, a firm's marketing strategy must set the direction and scope of the organization for the long-term that is consistent with the firm's business strategy. The marketing strategy should outline how the firm will achieve this through its configuration of resources, while considering the external environment, and meeting the needs of markets and to fulfill stakeholder expectations.

Strategic marketing is marketing that has been conceived and planned to leverage organizational strengths, while taking advantage of opportunities in the marketplace. Leading AEC firms use the marketing strategy as a framework to:

- Identify sources of competitive advantage
- Gain commitment to strategy

- Make strategic choices/decisions
- Provide a linkage to the company strategy
- Identify opportunities
- Align resources to invest in and grow the business
- Build processes and systems
- Inform stakeholders in the business
- Set objectives, strategy & metrics
- Measure performance
- Make a profit

The marketing function should be the strategic thought leader in an AEC firm. Marketing, whether an individual or a department, must look at the key drivers to of the firm's success and growth, and help align resources to deliver profitability and build firm value. If done correctly, marketing should provoke thought with your customers. Marketing is more than branding and creativity, it is about setting the direction and driving the firm to attract clients and keep business coming in the door.

To truly lead with marketing, a firm's marketing strategy must be an ongoing, dynamic process that enables a company to focus its resources on the right opportunities to increase profitability while satisfying the customer needs and achieving a sustainable competitive advantage.

The marketing strategy forms the foundation for the Marketing Plan, which consists of a series of initiatives and actions which meet marketing goals and objectives that help execute the overall business strategy.

The Marketing Plan itself should be developed on an annual basis and reviewed on a regular basis. The Marketing Plan is more than a document. If properly developed and implemented, the marketing efforts should link and unify all aspects of the organization. The Marketing Plan should include (see Appendix for Marketing Plan Basics):

- Overall strategy to market your firm and capabilities

- Situational analysis/SWOT analysis (see Appendix for background on SWOT)
- Market research, segmentation, and positioning
- Marketing strategies and tactics (marketing mix)
- Implementation plan (schedule and budget)
- Metrics and measurement plan (controls)

The Marketing Plan is based on a guiding set of goals and assumptions, and should not be developed and placed on the shelf. The Marketing Plan needs to be a dynamic tool to guide the marketing initiatives for the year. The Plan should also be subject to modification based on market conditions and changes in business strategy.

With the external and internal factors established, it is key to outline the following:

1. Your company's objectives;
2. The strategy you will use to accomplish these objectives; and
3. The tactics you'll employ to support the strategy.

This order is important. Too often, marketers start with the tactics, such as a brochure or new web site, and then try to match the objective or the strategy. Or, too often they simply implement tactics. All tactics should tie to the objective and strategy designed to meet business goals or they are simply an exercise.

Another key distinction that should be made is the difference between marketing and business planning. Simply, the marketing plan should evolve from-and support- the business plan and detail the strategy to help the business plan come to fruition.

Ultimately, the Marketing Plan must support the Business Plan objectives. The business planning process is the ideal starting point for ensuring marketing strategies and tactics are in line with business objectives.

Understanding Customers

"Customers buy for their reasons, not yours."- Orvel Ray Wilson

Before you can market—and eventually sell—to customers, you first have to understand exactly who those customers are. Understanding customers is critical to creating, delivering and managing customer value. Without this understanding of customer needs, it is impossible to develop successful marketing strategies. This understanding must transcend simple demographics like age, industry and geographical location to encompass the buyers' wants and needs, as well as their buying habits. This is also critical to uncovering opportunities you're your clients and target clients. Market research is a key part of defining your position in the marketplace, who your target customers are, and understanding what they want and desire. At the beginning of any marketing campaign, it is crucial to conduct market research that will provide you with answers to these all-important questions. After all, a good marketing program is built on solid market intelligence—in other words, cold, hard facts that will help reduce the risks associated with your campaign.

Grouping Customers

Once you've determined your overall customer base, you can begin to break it down in order to market more selectively to specific groups. Not all customers are created equal. They have different wants, needs, and willingness to pay for your services. Defining the market into segments allows you to group customers and target them with distinct marketing strategies. The market segmentation process divides customers into groups with similar, yet slightly different wants and needs. The basis of this segmentation can be drawn upon several lines: those who are seeking similar benefits from the service or product, how the product or service is used, project delivery methods, the frequency with which it is used, and the user's role in the buying process. Segmentation can also fall along more traditional lines such as demographics or markets.

Assigning customers to niches will break them down into even smaller groups with narrowly defined needs. This allows you to target this group with more specific messages and offerings. Although breaking down a group into such small niches takes considerable time and effort, the return on your investment tends to be much more substantial. Separating customers into niches also affords you the opportunity to differentiate between customers and potentially develop a closer relationship with each one.

"The aim of marketing is to know and understand the customer so well that the product or service fits him or her and sells itself." - Peter Drucker

Targeting Customers

Once you have a thorough understanding of your customers and their behavioral patterns, you can begin to target your marketing efforts. When selecting target audiences, you must define who is involved in the purchase decision. This will help you determine how best to develop and communicate the marketing message.

Your target audience should include both prospective and current customers, and should encompass both those who will buy your product or service, as well as those who influence the people doing the buying. In construction marketing environments, the person who is using the product or service may not necessarily be the one who buys or specifies it. Therefore, identifying and reaching the appropriate decision-maker can be a challenge, especially when there are several different buying authorities. This creates the need for contact at various levels within an organization. For example, a contractor offering commercial repair and maintenance services may focus their marketing efforts on maintenance managers, operation directors, buildings and grounds managers, reliability condo boards, owners, project managers, and facility managers.

In addition, the opportunity also may be controlled or influenced by a third-party. Typically, an engineering firm, architect or other consultant may have significant influence on a decision. Each decision influencer or decision maker may be looking for different things. Therefore, messages must be carefully crafted to their specific needs. In order to successfully market to an organization, making multiple contacts is necessary. However, keep in mind that this will result in increased marketing costs.

Customer Feedback

Obtaining customer feedback needs to be a continuous process. How else are you going to ensure that you are meeting the customers' needs? This is a critical step to ensuring repeat business and can be accomplished a number of ways. For longer-term projects, feedback on performance needs to be gathered on a frequent basis. At the conclusion of each project, send out a survey that solicits feedback about your performance on the project. Consider including questions about timing, safety, budget, communication, and any other factors that contributed to the completion of the project. There are several great online tools that can help you with this process including surveymonkey.com and zoomerang.com.

For larger projects, client debriefings are an excellent tool to get closer to the customer, and increase your proposal hit rate by better assessing target clients and developing strategies to satisfy their needs. In addition, debriefings are an excellent tool to demonstrate interest in clients, solicit feedback, and gather intelligence. Ultimately, debriefings lead to more work. Clients are typically very willing to share comments, both positive and negative. Either way, the feedback presents you with unique insights and an excellent learning opportunity.

Positive feedback is always great. When receiving positive feedback, send the clients a thank you note or e-mail and be sure to celebrate the success with your project team (give them a copy, post on the bulletin board and file a copy in the employee's file). With the clients' permission, consider using the comments on your web site, in brochures or case studies, as well as in a list of references.

While negative feedback doesn't feel as good, it provides an even greater opportunity. Any negative feedback warrants a call to the client to better understand their concerns and issues. This too provides an opportunity to address the issue in the organization and improve. Further, it demonstrates to the client that you sincerely care about them and you want to do a better job.

Starting with Research

There is simply no escaping the importance of a plan to guide your marketing efforts, though the planning process often stops when research is mentioned. Today, however, through the use of a myriad of online resources, the research component is finally manageable. In the marketing process, research is the foundation for effective decision making, and an essential step before segmenting, targeting and positioning.

Understanding Industries, Trends and Customers

Research is the collection, analysis and presentation of information. In AEC marketing, this information can be the basis for a wide variety of business decisions. While the role of the research function can vary by company, research can play a critical role in helping marketers and company leaders understand trends, dynamics, markets, customers, competitors and a variety of other information about their marketplace. Most important, this process helps uncover customers' needs and will serve as the basis for how your firm creates, delivers and manages customer value. Ultimately, research is the key to marketing successfully.

The primary goal of the market research effort will be to provide timely, accurate information about the company's prospects, customers, and the operating environment upon which decisions can be made. This information can be utilized to help management and the business development staff to make informed decisions. Marketing will be the central repository for information about the prospect and customer base as information can be obtained from a variety of sources and entails the attempt to capture information about the economy and specific business segments. This effort will allow marketers to track and analyze business indicators and be proactive in business and marketing efforts. Furthermore, the research will allow the ability to match the offering to the customers' and prospects' needs as well as introduce new products and services. Marketing research will also help to identify new opportunities for office locations, new products and services as well as new markets. The more you know and understand about your customer, prospects, competition and operating environment, the better prepared you will be to be successful.

Research helps guide your marketing efforts by answering questions that help you position your company to achieve success. By researching, segmenting and targeting your audience, you will be able to answer the questions: What are you marketing? To whom? Why will they buy it?

While most of us are not in a position to incorporate true primary research (defined as data gathered through first-person surveys and studies in an original format) into our marketing plans, secondary research (data gathered from other sources) has never been so easy to obtain. For example, most associations and publishing houses not only conduct industry research for members and subscribers, but often welcome suggestions or questions about an emerging trend, market conditions or industry predictions. Think of the power of sharing with your next potential client a trend you have learned. Services such as Hoover's and Datamonitor provide low cost access to industry and customer research.

Although search engines are a common part of our personal lives, we often forget to integrate them as a tool in our business research efforts. It is crucial to search for information on your competitors, clients, community and target markets at least once a quarter. Don't forget to also search for information about your own company, as this will demonstrate how your potential employees and clients view your company. With the proper data assembled, you can finally address the issues of predicted industry growth, target clients and anticipated trends for your marketing plan.

Customer Segmentation

"The mass market has split into ever-multiplying, ever-changing sets of micromarkets that demand a continually expanding range of options"- Alvin Toffler

Customer segmentation has long been a standard in the retail and automotive industries. There are countless examples of car manufacturers slapping a different label on an automobile and marketing it to a dissimilar demographic either based on a difference in price, image or even branding. The concept of customer segmentation also is standard operating procedure for the airline industry. Passengers with enough frequent flyer miles are part of an elite club that affords them the opportunity to board early, move up to first class or use a special lounge. Applying this same philosophy to clients in the design and construction industry, though, is a relatively new concept. However, it can prove extremely beneficial in terms of customer retention and profitability.

In its simplest form, *customer segmentation* is a process by which you organize your customers with regard to their needs and your services. Market segmentation is a process to categorize business targets based on a variety of criteria and factors that guide strategic and tactical decision making. A market can be defined as narrowly or as broadly as desired. The process allows a company to select and de-select customers to service, prioritize customers, define and deliver tailored value propositions for customer segments, as well as develop and manage effective channels for serving these customers. We all know that all clients are not created equal, but too often we provide the same level of service to our top clients that we provide to our least desirable ones. For example, while it is appropriate to work overtime and move other deadlines around to meet the needs of an "A" client — one that refers other profitable work to you, understands your value, does not question fees, provides challenging work and pays your invoices on time — it simply is not smart business to take that same approach for a one-time client that will never refer you to others, thinks you are too expensive and is slow to pay.

The first step in segmenting your customer-base is to identify and list the things you view as important in a client. Items on this list might include understanding the value you bring to the table, providing consistent referrals, acceptance of your fee structure, providing challenging work for which you are well-suited, possibility for a long-term relationship and future projects, upholding a true partnering spirit, is profitable and is a strategic fit in terms of market niche goals.

A primary goal of segmentation is to identify common similarities, differences, and characteristics amongst customers and potential customers. The segmentation process is required to make targeting and positioning decisions. Segmentation models can be built on a variety of factors including: need, industry, geography, SIC/NAICS code, size (revenue, employees, locations), purchasing processes, buyer behavior, project delivery method, and other factors. Segmentation allows companies to develop a deeper understanding of customer needs, and to better tailor their offering, marketing messages, and communication vehicles. In business-to-business marketing, segmentation can be a very effective tool, however, it is important to remember organizations do not make purchasing decisions, individuals make purchasing decisions.

With this list of desirable attributes complete, go through and rate your clients on a 1-to-10 scale in terms of how they perform as a client. Many find it simple to group the tallied numbers at the end in three categories — A, B and C clients. You will likely be surprised by the immediate realization that your company caters to many undesirable clients. Now, with this criteria established and a basis for where each client falls, determine what level of service is a fit for each category. This is often challenging for service companies to identify, but it's extremely necessary. In today's age, for example, many retail companies move C customers to web and phone orders only, ensuring that expensive time with a salesperson falls in the A or B range. They also break down ongoing service with regards to customer segments. Using these scenarios as an example, discern different offering levels for the products and services you provide. For example, an A client may require your top senior project manager, while a C client maybe be happy with a junior project engineer managing their project. Be sure the definition of service offerings goes beyond actual tangible deliverables to include areas such as communication. An A customer should warrant interaction from a senior staff member on a monthly basis, while a C customer should be part of a quarterly communication from the same staff member. The key is to be fair, not equal.

The end result from this planning process should be a customer action plan for your firm that guides your standard operating procedures for all clients. The segmentation process helps define specific marketing communication messages for each specific segment. While such tactics may seem out of the norm for your firm's culture today, most firms report that customer segmentation improves customer service since it enables better allocation and usage of your resources. Profitability also increases as a mechanism is established for getting paid for value delivered. But, most important, customer segmentation directly links customer behavior and potential profitability to customer loyalty — the key to success in today's highly competitive marketplace.

Jargon Junction

- **SIC:** Standard Industrial Classification code. Classification system used to classify businesses.
- **NAICS:** North American Industrial Classification System. NAICS is the standard used by Federal agencies when classifying businesses. In 1997, the NAICS code replaced the SIC code.

Establishing a Competitive Advantage, Differentiating and Positioning Your Company

Establishing a Competitive Advantage

Like many other industries, the AEC industry is very competitive. And competition is the basis for success or failure. More and more firms enter the market every day. Existing firms grow and diversify into new markets. Globalization is increasingly impacting different parts of the construction industry. To survive and thrive, a firm must establish a competitive advantage.

An AEC firm's competitive advantage can be derived from a number of sources, but ultimately is based on the value a firm brings to its customers. This advantage can be in the form of a unique combination of resources, a cost advantage, a market position, niche focus, financial resources, reputation, and experience.

In *Competitive Advantage*[2], Michael Porter, Harvard Professor and noted author, identified three generic types of competitive advantage as:

- **Cost leadership**- being the low cost provider of services or products
- **Differentiation**- being unique or different in an industry along a dimension that is desired by the customers. This differentiation can be based on product, service, experience, brand, location, or any number of factors. However, it cannot be independent of cost.
- **Focus**-this involves choosing to focus on a narrow scope of the industry

How does this apply to the AEC industry? It applies in a number of ways. Firms can choose to establish a competitive advantage based on one of the generic strategies, or combine them. There are many examples of AEC firms that focus on differentiation, serving a segment or specific niche in the market, or pursuing a cost advantage. Here are some examples:

- A ready mix supplier opened a batch plant in a major city where there will be significant construction. This helped establish a competitive advantage in terms of proximity and costs. (Cost Leadership)
- An engineer, contractor, and technology provider formed an alliance to offer design-build solutions for construction biomass plants. This resulted in cost and time savings for customers. (Cost Leadership and Differentiation)
- A contractor aligned with a material supplier to private label a unique specialty product and sells it directly to the customer. This resulted in lower cost and a turn key service for the customer. (Cost and Differentiation)
- A rebar supplier teamed with a post-tensioning supplier and offers a steel package for concrete project. This reduced the number of suppliers that concrete frame contractors had to deal with on projects. (Cost and Differentiation)
- An engineering firm may chose to focus exclusively on bridge design. This strategy included strategically hiring top bridge designers and former DOT personnel. (Focus)
- An architect chose to focus only on historic structures. This strategy involved hiring architects with significant historic experience and becoming involved in key trade groups. (Focus)
- A home building contractor chose to focus exclusively on timber frame homes. This included forming an alliance with a timber supplier and building the company's image in the market. (Focus)

Differentiating Your Offering

In marketing terms, differentiation is the process of distinguishing your product or service offing from your competition. Not all architecture or engineering firms are the same. Neither are ready mix companies, electrical contractors, steel erectors, or material and equipment distributers. Each firm is different, but not every firm capitalizes on their difference and translates it to meaningful value for the customer. There are a number of attributes about every company that makes if different from others. Part of the marketing process is to identify those differences and turn them into meaningful value propositions for the market. It is important to remember that the value proposition has to be of value to the market. This process is essential in marketing to help define the unique attributes of your business to create a sense of value. Simply put, differentiation is creating a benefit that customers perceive as being of greater value than what they can get from others. Differentiation, like the competitive advantage, is typically categorized in categories. This can include:

- Price
- Product/Service/Offering
- Performance Level
- Customer Service
- Focus/Niche

The objective of differentiation is positioning your company in the eyes of customers. The more effective your differentiation, the more easily you can compete on non-price factors. There are two primary sources of differentiation: cost advantage and differential advantage. Thus a competitive advantage can be achieved by delivering services or products at a lower cost to customers, or when the benefits exceed those being offered by the competition. These benefits can be in terms of resources or capabilities. For example, an architecture firm may have a competitive advantage in designing stadiums because they have successfully completed more stadium projects, and have a number of architects with stadium design experience. An engineer may have a competitive advantage in designing parking garages due to their experience and reputation. A material distributor may have access to lower cost suppliers that create a cost advantage. Ultimately, your firm is creating value for the customer through your competitive advantage.

Positioning

Positioning your company is an important element of the marketing process, and it aligns with your differentiation strategy. In fact, the decision on how to position your firm's offering of services and products needs to flow from your Business Plan. Positioning is the process by which companies create an image or identity in the minds of their target audience for a particular product, service or brand. A position is considered relative in comparison to competitive offerings. Positioning also represents a strategic choice. One company cannot be all things to all people. Positioning strategies flow down through the marketing mix and articulate how firms create and deliver value. Companies cannot be good at everything. If they try to, they typically are not superior in anything. Therefore, companies often set out to develop a value proposition, or ways they create and deliver value.

In *Positioning: The Battle for Your Mind* [3], authors Al Reis and Jack Trout describe positioning as "an organized system for finding a window in the mind. It is based on the concept that communication can only take place at the right time and under the right circumstances."

So, while position is something that happens in the customer's mind, marketing's role is to help influence that thinking. For firms that are selling something intangible such as engineering or architecture services, this can be challenging. Since services lack the physical attributes of products, you need to clearly articulate your value proposition to get to the customer.

Here are some common positioning categories found in the construction industry:

- **Low-cost:** essentially being the lowest cost option, based on price
- **Quality:** offering the best quality, but also at a high price
- **Performance:** being the most capable firm at delivering results
- **Service:** being the best provider of service associated with the sale

- **Variety:** offering the broadest array of solutions
- **Best value:** offering the best combination of quality and price
- **Innovator:** known for developing innovative products
- **Expert:** having the most experience or expertise in a given area
- **Leader:** being the largest, oldest, best known firm
- **Reliable:** being the most trusted, safest choice

While positioning is something that happens in the minds of the customer, it is the result of aggregate perception of customers in the marketplace. Ultimately, positioning should help the customer answer the question, why should I buy from you?

When attempting to differentiate and position your company, remember that you must be willing to sacrifice. To truly have a unique position, you must be narrowly focused. Reis and Trout stressed the need to choose one attribute that sticks in the minds of the buyer.

From a global perspective, positioning should help you understand how your strategies and tactics will help differentiate you and give your business a competitive advantage over your competitors.

SECTION 2: THE MARKETING MIX

Marketing: A Balance of Art and Science

"Authentic marketing is not the art of selling what you make but knowing what to make. It is the art of identifying and understanding customer needs and creating solutions that deliver satisfaction to the customers, profits to the producers and benefits for the stakeholders."

- Philip Kotler

As members of the AEC industry, we're all familiar with the elements that make up a great building: an inspiring façade and a sound structural system—in other words, a perfect marriage between art and science. What you may not realize is the same principle applies to your company's marketing efforts. To launch a successful marketing campaign, you must achieve a studied blend between art (creativity, design and strategy) and science (market research and measurements). Indeed, any marketing effort that attempts to rely solely on one or the other will be challenged to succeed.

The Importance of Good Design

Good design is the first crucial building block to a successful marketing plan. Why? As consumers, we're bombarded by so much marketing stimuli every day that our brains often take as little as a fraction of a second to decide whether to give something more than a passing glance — and in that time, what we notice are the visual devices a company uses to represent itself.

But how do you know what constitutes good design? The minute criteria are subjective but there are some common elements that design professionals will refer to again and again. The first is simplicity. Your message will never get through if it's hidden behind a bunch of clutter. Sometimes, especially if you have a very small space or a short period of time in which to get your point across, resist the urge to overstate, and the message will be that much more powerful. Think about Apple's use of stark white or Nike's short-but-sweet "Just do it" slogan,— each are great examples of how keeping things simple can work in a company's favor.

It's tempting to jump on the bandwagon and follow whatever design trend happens to be hip at the moment, but to do so would be to ignore another important tenet of good design. It must relate back to your business and your core audience. Often, this is done in extremely subtle ways, through font or color choices. The use of green in Starbucks' logo, for example, reinforces the company's commitment to environmentally-friendly, fair-trade products. If you're working with a designer to create new marketing materials, it is important he or she understands the nature of your business and can translate that into the design elements.

The final, and possibly most important, component of good design is that it be consistent. Our brains are wired to recognize patterns, and thereby we're more likely to remember repeated elements. The consistent use of fonts, colors, logos and slogans is all part of the larger process of branding. Creating a recognizable brand should be a goal of any marketing campaign, as it is this recognition that will separate your company from the rest of the pack. To achieve this impact, the brand image must be consistently delivered through all marketing mediums.

Backing It Up With Science

Ultimately, marketing is about understanding and satisfying customers' needs. So it stands to reason that the best-designed marketing campaign in the world is useless if it doesn't reach its intended audience and deliver the desired result. That's where the other half of the equation — science — comes in. Science is necessary at the beginning of the campaign to identify, define and target markets. The research component also includes understanding prospective customers' needs. In addition, science is also just as crucial after the fact, when measuring customer response and Return On Investment (ROI). Ultimately, achievement of stated marketing and business goals will help you determine the effectiveness of your campaign.

Common sense dictates you must know whom you're selling to, and what is important to them, before you can sell a product. Therefore, most of us need little convincing about the importance of conducting market research before embarking on a marketing campaign. But, with today's savvy business buyers, a general overview of the market is no longer sufficient. Instead, market research is moving toward segmentation — specific, targeted research that identifies subsets of the whole group in order to provide more specialized marketing efforts.

Research also should extend beyond just who your customers are to address how, why and when they buy the things they do. These variables, particularly the latter, can help dictate another important, scientifically-minded aspect of marketing: timing. For instance, if you know that many businesses review their budgets and make decisions about new construction projects at the end of the fiscal year in July, you'll have a launching pad for how and when to approach these potential customers.

No matter how much research you do on the front end though, you can never really gauge the success of a marketing campaign until it's been presented to your target audience. But, getting the campaign out the door is only half of the job. In order to ensure future success, you'll also need to measure its impact. This starts with benchmarking, to ensure there is a standard to which you are measuring against, and the establishment of measurable objectives. In marketing terminology, these measurements are known as metrics, and analyzing them can help you reach the ultimate measure of success: return on marketing investment, or ROMI. Traditional marketing metrics have included data focused on tactical actions, such as impressions, reach, response rates, web site traffic, click-throughs and search engine rankings. However, the emergence of electronic communication and social media has created new challenges for marketers. Effective scientific measurement requires focus on the full range of marketing impacts, which, in addition to traditional measures, also includes factors such as changes in market position, increases in brand perception and value, and achievement of overall marketing goals and objectives. Ultimately, in business-to-business marketing, the bottom-line return matters.

Just like great buildings, marketing campaigns must depend equally on art and science if they are to be successful. The art of marketing helps make the science of marketing complete. Take one component out of the equation, and you could find yourself building your company's marketing identity on a foundation that's less than solid.

Marketing Mix

Traditional marketing textbooks define the Marketing Mix as the "Four Ps" of marketing: Product (Service), Price, Place (of distribution) and Promotion – all of which are used to describe the strategic position of a product or service in the marketplace. In *Marketing Insights From A to Z* [4], Phillip Kotler described the marketing mix as "the set of tools management uses to influence sales."

Marketing's role within an organization is to optimize the marketing mix. Marketing effectiveness and efficiency is achieved by offering the product with the right combination of the four Ps of marketing. Constant monitoring of marketing results in tactical changes involving changes to the different elements of the marketing mix. There are a variety of books and resources that offer more details of the Four Ps of Marketing, but the intention here is to provide a brief overview.

- *Product (service)*: This describes the product or service offering, including characteristics and attributes of a product or service. Differentiation of product can be based on a variety of factors.
- *Price:* This defines the cost to the customer in acquiring your product or service. The price is determined by several factors including costs, competition, product and brand identity and the customer's perceived value of the product/service. This includes the initial and on-going costs associated with your product or service.
- *Place (of distribution):* Where and how customers can acquire your products/services. This is typically referred to as a distribution channel.
- *Promotion:* Refers to a combination of activities to promote product and services including advertising, direct response, PR & publicity, sales efforts, and other efforts.

For each company, there is a different marketing mix. The marketing challenge is to integrate and manage the mix that delivers results for your company.

Marketing Messaging

"If you're trying to persuade people to do something, or buy something, it seems to me you should use their language, the language in which they think." - David Ogilvy

A well-developed marketing message has been the backbone of branding since the dawn of market strategy. Given the rapid pace of technology today, the marketing message has taken on new forms and is engaging with audiences at an unheard-of rate. With advances in technology, such as the Internet, mobile technology and visual media, it no longer matters if you are smack-dab in the middle of Times Square in Manhattan or nestled in a quaint village in the heartland — a clear marketing message can have a far-reaching impact, especially in the construction industry.

The Importance of a Message

As society embraces new technology, a person's attention span continues to decrease, which makes the development of an effective marketing message essential to building your brand and implementing your marketing strategy. Considering that a potential client receives and views thousands of messages from multiple sources each day, your message has to be compelling and leave a lasting impression.

But it's not enough just to create a strong message — you must consistently position and brand the message in order to achieve the desired impact. Create a plan to properly reinforce the message throughout your organization, whether it is a tag at the end of an e-mail or on a job site sign. All of your external media outreach should bear the same message, and its design and delivery should be consistent.

Defining a Target Audience

In order for a message to really get through, it must be clear and concise and offer a compelling reason why someone should do business with your firm. Therefore, you should have a detailed understanding of your target audience and their perceptions and needs before you begin to develop your message. To do this, you should ask yourself the following questions about your target audience:

- Why do they want/need your product or service?
- What do they know about your product or service?
- What are the alternatives, and what do they know about them?
- What is most important to them?
- How will they buy? Decide to buy?
- What questions will they ask?
- How does your message address these?

Once you've determined these answers, you can start to create a message that demonstrates how your product or service meets your target audience's needs.

Challenges

Stating how you will meet your customer's needs is only half of the marketing message equation, though — you'll also need to differentiate yourself from your competitors. Differentiation can be based on features, benefits, functionality, price and/or other factors, and it enables the prospect to make split-second decisions. To determine what sets you apart from your competitors in the eyes of your target audience, you'll need to decipher their points of pain or nuisance.

By defining ways in which the majority of your audience is tied down to certain restrictions, you can craft your message to meet a centralized demand. If you solve a nagging problem or offer an alternative for a prospect, chances are you can turn them into a customer for life. Ask yourself the following questions to determine challenges prospects face:

- What keeps your prospects up at night?
- What problems can you solve?
- How can you improve something they are doing?
- Do you offer an alternative?
- What are their needs, wants, pains and fears?

- How can you articulate that you understand their problem?

Presenting the Solution

After you have identified these potential challenges, the next step is to offer a solution to them. The messaging process must cross the chasm from being about the product or service to being about the customers and their needs. One method for demonstrating your success in solving customers' problems is through the use of testimonials and case studies — in other words, examples of how your solutions have helped other clients. This helps provide credibility and generates a feeling of confidence that your product or service has performed for others. A prospect is likely to empathize with relative case studies and testimonials.

While drafting testimonials, use keywords that jump out at your target audience and reinforce your message. Be sure to select users of your product or service who are easily noticed by your audience and predominant in your organization's market. Providing a testimonial from a company that your audience has never heard of could impact the effectiveness of a message.

Building the Brand

"Your premium brand had better be delivering something special, or it's not going to get the business."- Warren Buffett

After you have a thorough understanding of who your audience is and what troubles them, you are ready to begin crafting your message. Keep in mind that the overall message and value proposition must be consistent with your brand strategy, but can be modified for specific audience segments. Be sure the message addresses solutions for the target audience's problems, as well as the following:

- Why should they do business with you? What is your value proposition?
- What does your brand mean?
- What expectations, performances and associations does your brand evoke?

- What makes you qualified? What are your past experiences and track record?
- Why should they trust you more than others?
- What is the promise? How is it supported? Is it credible?
- Is it unique? How does it differentiate you from the competition?
- What is in it for the prospect?

An effective message is one that is simple to communicate and simple to understand — one that will clearly articulate your positioning and make a positive and memorable impression. Once the message has been developed, it is time to test it. Before releasing it to your prospective clients, consider creating a focus group that can evaluate the message. Use the suggestions from the focus group to refine your message and ensure it hits home with your prospects.

Once the message is finalized, it's time to get it out to your audience. Start by incorporating it into all external communication. Make sure the message is used consistently through all advertising, collateral, web site and mail pieces, as well as any other marketing communication vehicles. You'll know your message has made an impact when more leads are generated and sales increase. Building a brand isn't a quick or easy task, but it is well worth the time and effort you'll invest in doing it.

Value Proposition

The customer value proposition consists of the sum of the benefits and value derived in return for payment of the product or service. Simply stated, the value proposition is what the customer gets for their payment. For AEC firms, the value proposition must move beyond the specific features of a product or service and be articulated in terms of the benefits and values received by the customer. How do you create and deliver value to the customer?

From a marketing and sales perspective, you need to craft a value proposition that doesn't speak to your product or service's feature(s), but focuses on the benefits derived from that product or service and the value created for the customer. A simple, but effective illustration of this is grass seed. People don't buy grass seed because they want grass seed. They buy grass seed because they want a lawn. The marketing challenge is to look at the product or service through the eyes of the customer or potential customer and translate the product or service features to benefits and communicate value.

Grass Seed (Features)

- What is your grass seed?
- What are the features of your product or service?
- What are the unique attributes?
- How is the product or service different from the competitive offerings?
- How does the price reflect the value?

Lawn (Benefits and Value)

- What is your customer's lawn?
- What is the customer buying?
- What problem are they trying to solve? What pain are they trying to alleviate?
- What is most important to the customer or perspective customer?

- What key factors do they value the most?

The value proposition is an essential element of an AEC firm's marketing and sales strategy. For example, the value proposition may provide a guide for a firm to target a specific market, help contribute to a differentiation or positioning strategy, be central to a marketing communications strategy, and/or be part of the sales communication process. Most important, the fulfillment (or delivery of the product or service) must be consistent with the expectations set forth in the value proposition.

Jargon Junction

- **Value Proposition:** sum of benefits and value derived in return for a specific product or service
- **Feature:** specific attribute of the product or service
- **Benefit:** what the customer gets when acquiring the product or service
- **Value:** customers' expectations of value relative to the price paid to acquire the product or service

Building the Brand

For many AEC firms, the company's brand is one of the most valuable company assets. A brand is reflective of the value of the products and services generated in the minds of the people. A brand defines the company, product or service offering in terms of specific characteristics and attributes. These attributes include logos, slogans, designs, messages and emotions. You must build perception of value and consistently deliver to that perception. The brand value embodies economic and non-economic, tangible and intangible factors. Brand value is a direct result of the sum of the marketing communications efforts and activities, as well as the customer's experiences with the company, product or service. When making purchase decisions, people take into account a wide range of factors including features and benefits, product or service personality, identity, image and value. How a brand is known and understood is key to success.

A brand development process begins with developing a brand concept. A key principle in building a brand is having a clear understanding of customers and delivering benefits they truly value. Establishing and building a strong brand is essential to gaining acceptance, preference and eventually raising prices. All brands are built around the "Four Ps" -- product/service, price, promotion and positioning/place. If value is not perceived by the customer and a product or service has not been properly differentiated in the mind of the customer, then it is a commodity; therefore, competition is based on price.

In *Positioning: The Battle For Your Mind* [5], Reis and Trout asserted that a company must create a position in the "prospects mind, a position that takes into consideration not only a company's own strengths and weakness, but those of its competitors as well."

Positioning the brand involves marketing communication initiatives to position the brand relative to competitive offerings. Typically, this is segmented based on factors that the consumers value. What can the brand do for them? Branding is a promise of value to the customer. If the brand promise is not delivered, brand value is diminished. A key to branding success is articulating the brand value and communicating why the customer should buy.

Once the brand is established, it must be managed. For AEC firms, brand management involves a variety of marketing techniques to enhance and increase a brand's value in the minds of customers and potential customers. Over time, brand equity is built. Brand equity is the value of the brand's name, logos and symbols, identity and reputation. The brand value is the financial perception of the value of the brand. Thus, developing a strong and consistent message and image for your company is critical. These efforts result in building and positioning the brand. Eventually, this will lead to brand loyalty and allow the company to maximize sales and maintain a strong position in the marketplace.

A strong brand is important because:

- The company (products and services) will be recognizable within the industry
- The brand sets the experiential expectation
- The company will be known for certain attributes (quality, service, price, etc.)
- Customers can expect a certain level of quality and performance
- Price comparisons are reduced and less important to prospects and customers
- Prospects and customers will associate less risk with a known company
- Customers and prospects will be familiar with your company's capabilities
- Customers will develop a preference for your company
- A strong brand will have positive associations
- A strong brand will eventually lead to higher prices

Branding Checklist

Elements of brands are a distinct bit of visual, verbal and emotional information that identifies and differentiates a product or service. The brand is established to create awareness, associations and set expectations for customers that choose a product or service.

Brands should be Unique/Memorable/Likable

- Is it easy to pronounce?
- Is it easy to recognize and remember?
- Is the brand unique and distinctive?
- Does it attract attention?
- Is the brand appealing to the customer?
- Does the brand evolve positive emotions?
- Does it standout in a group of brands?

Brands should be Relevant

- Does it suggest product/service attributes or benefits?
- Does the brand help describe specific elements or attributes of the product or service?
- Does it distinguish the product relative to the competitive offering?
- How can the brand tie into the brand positioning strategy?

Brands should be Expandable/Flexible/Available

- Can the brand be extended to other products or services?
- Can the brand be used in different languages, geographies and cultures?
- Can the brand support positive associations?
- Can the brand be updated and refined for future applications?
- Is the brand name and visual identity available?

- Can the brand name and visual identity by protected by trademarks?
- Are appropriate web domains available?

Brand Management Checklist

Developing a brand identity for your firm can be challenging, and managing it can be even more challenging. The brand must tie to the 4Ps of Marketing (Product/Service, Price, Promotion, Positioning). We've prepared the following Brand Planning Guide to assist in this process.

Product/Service:

- Is the product/service aligned with the brand's vision?
- Does the brand evoke positive emotions with your customer/ prospect base?
- Can the brand be extended to cover complimentary products and services?
- Will the brand drive customer loyalty?

Price:

- Where do customers value our brand in relation to competitive offerings?
- Can the brand be priced at a premium compared to competitive offerings?
- How do price perceptions impact our brand?

Promotion:

- Are we consistently representing the brand throughout all company marketing and communication?
- Are we using an integrated marketing communications strategy to promote our brand, products and services?
- Do our employees understand our brand vision and value propositions?

- Do we have a brand standards guide?
- Is there a single person or department responsible to ensure brand consistency?
- What metrics and tools are in place to measure brand value?

Positioning/Place:

- Does our brand strategy articulate how our product/service offering meets the needs of our customers and prospective customers?
- Does our brand strategy differentiate; position the company, and motivate customers and prospective customers to choose our brand?
- Can our brand be applied across multiple segments?

Once brand presence is established, companies must strive to maintain it. Consistency in execution of marketing communications is critical to maintaining the brand. Building a brand does have its risks. It only takes one bad experience to impact a customer's perception of a brand.

Marketing communications programs focus on communicating the brand externally to various markets. An important, but often overlooked component of branding is internal branding. Company employees must understand the brand and value proposition. Not only must they understand the value proposition, they must embrace it and carry it through to the customer.

Establishment of a brand is a significant effort and investment. Developing a logo and brand identity is an important first step. To ensure consistent use of the company brand, invest in developing a corporate identity manual. Such a manual specifies how the elements of a company's identity should be used. This effort will ensure consistent use of the identity throughout the organization and by outside vendors. The manual should specify identity designs, positioning, corporate colors, fonts, layouts and taglines and acts as a guideline for everything including advertising, stationary, signage, clothing and packaging.

Brand identity should not be confused with corporate identity. A corporate identity is essentially how the company is perceived by external stakeholders, such as, prospects, customers, vendors, government and potential employees. In addition, corporate identity can also have an impact on internal stakeholders, such as employees, owners, investors and key partners.

Corporate identity is influenced by controllable and uncontrollable factors. Marketing can play a large part in shaping the corporate identity, but marketing should not be used alone. Every interaction someone has with the company has a potential impact on the corporate identity. A corporate identity should be rooted in the company culture and must be planned, crafted and managed.

Jargon Junction

- **Brand:** A brand consists of a combination of attributes, tangible and intangible, symbolized in a word or visual identity that creates value and influence. A brand evokes associations, expectations, and performances.
- **Brand Associations:** These are emotions, feelings and an understanding that customers and potential customers have about a brand. These associations can be built through marketing communications, but are also the result of ones experience with the brand.
- **Brand Identity:** This includes the name and visual representation of the brand. The brand identity is the foundation of any branding effort. The brand must be relevant, recognizable and help differentiate the offering.
- **Brand Management**: This includes all efforts to manage the tangible and intangible aspects of the brand, including the "four Ps"-product/services, price, promotion, place/ positioning.
- **Brand Positioning**: This is the distinctive position that a brand adopts relative to competitive offerings. Consistent execution is required in every element of the marketing mix.

- **Brand Strategy:** A systematic plan to develop a brand so it meets its objectives. The strategy should be driven by the points of differentiation and positioning. It should be consistently executed through all integrated marketing communication efforts.
- **Branding:** The process of identifying and blending tangible and intangible attributes to differentiate a company, product or service in a compelling manner.
- **On-line Reputation:** A summation of a business or personal brand or reputation through a combination of electronic media, such as blogs, social networking sites and other digital sources.
- **On-Line Reputation Management (ORM):** A systematic process of monitoring an online brand or reputation across online media.

Creative Development

Whether through direct response or advertising initiatives, the creative execution of your marketing initiatives plays a large role in your firm's success. AEC marketers have more tools than ever available to them. Computer-generated graphics, renderings and 3D models are interesting and engaging but need to be part of an overall creative plan. Marketing materials and campaigns must effectively communicate messages, differentiate and position, build awareness and compel the target audience to take action. The most creative marketing tools may be completely ineffective if they don't achieve their prescribed goals. Many companies have a formal creative development process, which takes a variety of factors into consideration including: goals and objectives, messages, marketing forum, desired actions, etc. When you are developing creative materials, consider utilizing an advertising agency or outside creative resource. Outside sources can bring fresh ideas and perspectives, but this must be carefully balanced with the marketing objectives in mind.

Creative and Campaign Briefs

The creative process begins with the strategy development represented by the campaign and creative briefs. Creative and campaign briefs are effective tools for developing marketing campaigns or individual collateral, advertisements, direct response pieces or web sites.

Creative ideas must be consistent with the marketing and brand objectives. Too often, creative for creative sake does more harm than good. Simply, creative must also be consistent across all marketing efforts. Not the same, but consistent. Creative should help build the brand identity of your firm. Ultimately, your prospects and customers should know your brochures, ads, collateral and other marketing material when they see it. And remember, your creative material should help articulate how your firm creates and delivers value for the customers.

Creative campaign briefs should provide a good overview of target audience, messaging, benefits, features and value statements.

Following are some key tips and terminology to review with as you begin developing your marketing creative campaign:

Purpose: A succinct statement of the purpose of the campaign and expected outcomes.

Target Audience: The target audience can be described by markets, company, position, function or role. Keep in mind that there are likely to be multiple decision-makers and influencers. Each should be listed and addressed. Consider rational and emotional influences on their decision.

Objective: State the objective of the overall campaign or creative piece. An ad campaign to increase brand awareness will take a significantly different approach than a direct response program designed to elicit a response. The objective should address the prospect's points of pain.

Perception: *Current Perception-* What is the current perception of the target audience about your product or service? *Desired Perception-* What perception do you want the target audience to have? How are you going to communicate this?

Pain Points: Needs, wants, desires, concerns, issues and hot-buttons of target audience.

Promise/Message: What is the promise? What is the key message you are trying to communicate? What are you offering the target audience? This must address the audience's needs and address your objective.

Support: How will the promise/message be supported? What credible information supports the promise/message? How can we make this message/offer compelling?

Approach/Tone: What is the desired tonality of the creative piece? How will the audience be addressed? How will the piece look and feel?

Desired Actions: The call to action could be to visit the company web site, contact the company via phone, visit the trade show booth, request a white paper, etc.

Mandatories: Items that must be included. Examples are web site address, phone numbers, faxes, colors, logos, etc.

Measurement: How will the results be measured? What are the goals of the effort? What are the benchmarks or indication of success?

Integrated Marketing Communications

"Marketing is not an event, but a process . . . It has a beginning, a middle, but never an end, for it is a process. You improve it, perfect it, change it, even pause it. But you never stop it completely."- Jay Conrad Levinson

When asked to define "marketing," many might struggle to come up with an answer. That's because there's not just one thing that encapsulates marketing as a whole; rather, the term refers to anything a company does to create awareness, interest and demand for its products and services. Marketing communications is designed to deliver a clear, compelling message to the right decision-maker or influencer. However, in today's world, there are more and more ways to communicate this message than ever before, and therefore marketing efforts must continue to evolve to more effectively build brand awareness and preference while also delivering a measurable return on investment. That's why savvy marketers are embracing the concept of "integrated marketing communications" as a systematic way to coordinate marketing efforts across all possible channels.

Integrated marketing communications is a strategic approach designed to harness all aspects of marketing communication such as advertising, direct marketing, public relations, sales promotion, e-marketing, collateral and other programs to work in coordination as a unified force. This approach enables the consistent delivery of coherent brand positioning through all marketing communications mediums. An integrated effort consists of planning, coordinating and controlling the marketing communications process with the end result being a synergistic, seamless, customer-focused marketing program.

The integrated marketing communications concept particularly applies to professional services firms, where everything communicates and creates an impression with prospects and clients.

The development and implementation of an integrated marketing communications program involves four steps:

1. Develop a Database

At the foundation of any integrated marketing communications program is a robust database. This serves as a data warehouse with information about your customers and prospects. The database should be integrated with a customer relationship management (CRM) or sales force automation (SFA) system, and should track information about prospects, customers, decision-makers and influencers. This database will allow a marketing team to measure and evaluate customer purchase behavior and gain valuable customer insights, which will in turn help identify the target markets and segments crucial to developing a marketing communications strategy.

2. Strategic Marketing Plan Development

The strategic marketing planning process should involve internal and external research and analysis, market segment analysis, marketing plan development, resource allocation and budget development. A core brand strategy must be developed to ensure that brand communications are clear and consistent through all aspects of the customer's experience with a company. A core tenet of this strategy is that all communications emanate from a single strategic marketing platform.

Integration of all messages will maximize reach, power and impact. Integrating each component of the marketing program will leverage and draw from the strength of the other initiatives. For example, direct mail will more likely be opened because the recipient saw an advertisement, a magazine reader will be more likely to respond to an ad because they read an article about a product or service, a trade show attendee is more likely to approach personnel at the show because they received an e-mail or have heard of the company before and a web surfer will be more likely to visit a web site because a search engine ranks a site high in its search lists.

Another key component of integrated marketing communications is the focus on multiple stakeholders. Traditional marketing approaches may only address one stakeholder or target audience— usually the customer. Integrated marketing communications, on the other hand, integrates communications across multiple stakeholders including customers, prospects, current and potential employees, vendors, investors and media and business partners.

3. Tactical Implementation

Once the marketing plan has been developed, it is critical to build a structure for implementing, managing and executing the marketing program. Implementing an Integrated Marketing Communications (IMC) program requires careful orchestration of all marketing activities, continual monitoring and articulation of a clear and compelling message.

The ultimate value of an IMC is synergy. The synergistic effects of integrating marketing initiatives results in a sum that is greater than its parts. By leveraging and maximizing the customer contact points, IMC results in:

- Creative integrity
- Consistent messages
- More efficient use of media
- Cost savings
- Noticeable results

Ultimately, the IMC pulls together all elements of the promotional mix across all contact points. While it may seem difficult to achieve integration, the risks of not integrating marketing efforts can have several impacts including lack of coordination of messaging, as well as communication and marketing vehicles that can result in inconsistent, contradictory and unconnected efforts.

4. Measurement and ROI

The final component of an IMC program is measurement. Any marketing program must have specific and measurable goals and objectives that are not just tied to leads and awareness, but also encompass things like proposal opportunities and profitable jobs. Focus on measuring these goals and generating ROMI will lead to continual refinement and improvement.

Companies seeking an efficient, cost-effective, controllable and measurable marketing program must commit to an IMC program. IMC is more than the coordination of initiatives; it is an aggressive, dynamic and results-oriented marketing plan that can deliver results. Those companies that implement IMC programs will enjoy a significantly greater return on investment than those who rely on traditional independent marketing efforts.

Components of IMCs:

- Research
- Market segmentation and planning
- Messaging and positioning
- Implementation of tactics
- Measurement and control

SECTION 3: MARKETING TOOLS

Advertising

"You must have mindshare before you can have marketshare." - Christopher M. Knight

We see them every day on television or on the side of the highway and while flipping through our favorite magazine or newspaper. Occasionally, we find one that catches our eye and raises our level of interest in a product or service. There is little argument that advertising is an important component of building a company's brand. It reinforces all of the essential elements of a defined public relations campaign and distinguishes firms from the competition. But how do you, as a member of the construction industry, go about making decisions on where to advertise?

Advertising Defined

To begin, it is important to clearly understand what advertising is and is not. A common mistake is using marketing, advertising and public relations as synonyms. Advertising is merely one component of a complete marketing plan that mixes various communication tools, mediums and processes to best communicate your message. According to *Barron's Dictionary of Marketing Terms* [6], advertising is the "paid form of a non-personal message communicated through various media. [It] is persuasive, informational and designed to influence the purchasing behavior and/or thought patterns of the audience." In other words, advertising (television, radio, print or web) should inform potential customers about your product or service, its specific attributes and how to get it. In *Marketing Insights From A to Z* [7], Phillip Kotler described the "mission of advertising as being one of four: to inform, persuade, remind or reinforce a purchase decision." Advertising is directly tied to sales in that it educates potential consumers about a product or service you have approached them directly about.

During the last few years, the style and role of advertising has changed considerably. We are now bombarded with so many advertising messages every day that our brain takes mere seconds to sort out which messages we will remember. Today, consumers can have instant access to a product and have therefore come to expect to be directed to a relevant medium.

What hasn't changed over the years is the importance of advertising within the overall concept of brand marketing. According to Al and Laura Ries in the *22 Immutable Laws of Branding* [8], "Advertising is a powerful tool, not to build leadership of a fledgling brand, but to maintain that leadership once it is obtained. Companies that want to protect their well-established brands should not hesitate to use massive advertising programs to smother the competition." So why is it that so many companies in the construction industry claim that they have tried advertising and it failed? Chances are because their programs lacked research, planning and consistency.

Planning the Campaign

No different than any construction project, planning is the first and most crucial step in ensuring advertising success. Too often, the appeal of the last-minute deal in a publication forces concrete professionals into haphazardly developing an ad that does not accurately convey their message. Developing a proper plan will require you to research which media outlets (television, trade magazines or newspapers, Internet or radio) are the best fit for your company. Ideally, your selection should be based on your ability to reach your target audience in the most cost-effective manner possible. For the most effective coverage, you could choose to go with multiple media vehicles with an integrated marketing communications effort.

However, this step is simply the beginning. Once you have selected the ideal outlet for your ad placement, it is important that you carefully consider the timing of your message. For example, if you've decided to place an ad in a trade publication, make sure you review that publication's editorial calendar so you can determine which issues are most relevant. Most publications will be happy to supply you with an editorial calendar, part of a larger media kit that includes standard ad rates and demographic information, when you call to inquire about advertising. Without consideration of outlet and timing, you could be in danger of reaching the wrong audience.

The next step is to develop a goal for your advertising efforts. Is the goal to generate sales or increase your brand? This will help you clearly define your message in the ad. This message should be consistent with what you use in all external marketing communications. It should also communicate distinctive value or attributes of your company, product or service. However, within the context of an advertisement, it is important to develop a call to action, so readers know what you want them to do. For example, if the goal of your ad is to generate sales for a new product, make sure to tell readers where they can go to learn more about it or buy it. It is important to remember an ad simply displaying your logo and company information probably won't garner phone calls and leads, though such a piece can serve to brand your company. However, an ad offering a free report or unique information may result in contact.

Design and Timing

Readers will decide in a split second whether to keep reading your ad. Therefore, the graphic layout of your ad must be visually appealing. Do not clutter your message or overwhelm the reader with massive amounts of copy. Further, make sure to document any claims you make and avoid phrases that cannot be substantiated such as "We are the best." The goal of any ad is to drive the reader to perform an action — such as remembering your company the next time they need your service or product. The copy should address the needs of the reader first and foremost.

Frequency

One of the key elements of a successful ad campaign is frequency. If you chose to run a full-page ad one month, but your budget does not allow you to run such a large ad every month, you should consider running smaller ads on a more frequent basis. Repetition will reinforce your message and provide the greatest potential for success. Advertising is not a one-time event, but a process. When just starting an ad campaign, many companies may give up after only a few ad insertions. Research indicates a person needs to see an ad five to seven times before responding. Be sure to build a relationship with the advertising salesperson as well, as they are often in a position to create a custom program to better meet your needs. A partnering spirit will go a long way with a publication and afford you the opportunity to build a program that goes beyond advertising to sponsorship activities, directory listings and other special venues. Also note many trade publications have an online presence as well, such as online magazines, e-newsletters, webinars, etc.

Advertising Vehicles

"We find that advertising works the way the grass grows. You can never see it, but every week you have to mow the lawn." - Andy Travis

A chapter on advertising would be incomplete without discussion of the different types of advertising vehicles available to those in the AEC community. Media should be selected based on its ability to effectively reach the target audience.

Trade Publications: There are several hundred trade publications which target very broad and very specific parts of the AEC community. The variety and diversity of publications can help marketers select the right media vehicles to deliver a specific message to the audience. When reviewing trade publications, consider the audience, editorial mission and association affiliations of a publication. For example, in the concrete industry, *Concrete International* is published by the American Concrete Institute and offers highly technical articles for architects, engineers, contractors, researchers, manufacturers and technicians. *Concrete Construction*, which is published by Hanley Wood, is geared to the concrete contractors and focuses on concrete construction materials, methods and products. A marketer may choose to use both publications, but it is important to understand the audience and their purposes for reading the publications.

Online Avenues: The growth of the Internet has given birth to a wide variety of new advertising forums to reach prospects and customers. Moreover, technology has enabled more interactive advertising. AEC marketers have a variety of online advertising options ranging from banner ads, social networking advertising, e-mail advertising, site sponsorships, search engine marketing, and video marketing. Many publications offer online versions of their publications, ezines and specific portals. Internet advertising has led to an important change in the dynamics of marketing: responses are immediate and conversion to business is very high.

Directories: Another option that can take many forms is directories. Many associations and publications publish directories. One example is the *Blue Book of Building and Construction*. The *Blue Book* is a series of regional construction directories published annually that list providers of construction, contact information, products and services. In addition to the listings, the *Blue Book* offers paid, display advertising. They also offer an online version with searchable databases at www.thebluebook.com. The Blue Book is a particularly strong tool for product suppliers and sub-contractors, but not as effective for engineering firms. For AEC firms, directories are an effective way for finding providers of products and services, and therefore are a good opportunity for marketers.

Advertorials: Many publications offer advertorials, which are a combination of advertising and editorial content. Under this concept, firms buy an ad and also get accompanying space to submit content about their products, services or projects. The use of advertorials is increasing as publications and web sites seek ways to increase revenue. While the impact and credibility of advertorials may not be as good as traditional advertising or public relations, they can be effective. Publications such as *Engineering News Record* (ENR) have been offering special sections that contain a combination of advertising and editorial for years. There also has been an increase in the number of vanity publications that are produced to feature specific construction projects, cities and companies. Produced by custom publishers, these publications can be effective means of communication, but do your homework first. Investigate the publisher, understand the editorial mission, and discuss the circulation and distribution before committing to advertise. These custom publications are very profitable for the publisher, but may not be the best investment. There has also been a rise in special television and radio programs that offer to do a feature on your company if you agree to advertise or underwrite the costs.

Audience/Reach

Regardless of the publication you choose, it is key to understand who it reaches. The Business Periodical Association Worldwide (BPA) is an independent entity that audits the circulation of publications to verify and qualify circulation. Qualified circulation is an important factor as it represents a publication with proven circulation by documentary evidence, meeting the market serviced as set forth by the publisher. The publication's circulation can be paid, non-paid, or some combination of both. Through this auditing process, a variety of demographic information is collected about the readers of the publications. This information is very useful for marketers to identify and target the right audience. In recent years, BPA has expanded its offering to track web and new media viewership and readership. The BPA statements provide insight about subscriber demographics, circulation trends, price, frequency, and affiliations with associations. Most publishers include BPA Statements in their media kits or on-line. The BPA also offers a variety of other information for media buyers. For more information, visit www.bpaww.com

Another source of business periodical information is Standard Rate and Data Service (SRDS). SRDS offer information on publications that provide specific information such as standard rates, ad specifications, contact information and links to on-line media kits for business publications by market and industry. The SRDS publications provide a summary of media kits for publications. In addition, they offer publications that offer detailed information about direct marketing lists. For more information, visit www.srds.com

Audited publications are those with a controlled and verified circulation list. This means that subscribers are qualified to receive the publication, or they must meet a certain criteria for receiving it. From a marketer's perspective, the audited publication offers a level of validation, integrity, and relevancy to the media planning process. For example, to receive a certain non-paid engineering publication, you must certify you are an engineer and provide some demographic information.

In contrast, vanity publications are typically produced by a publishing company and distributed to a broader audience (often times at an author or organization's expense). Research has indicated readership in vanity publications is not as high as in audited publications. Several publishing firms have been publishing "best of" construction books, or special issues promoting a firm or project. In many cases, these firms request supplier lists and then seek ads from suppliers to underwrite the publishing costs.

Measuring Success

"Half my advertising is wasted, I just don't know which half." –Jonathan Wanamaker

Once your advertising message is out there, how can you tell if it is effective? The simplest answer is you'll see an increase in leads and other inquiries. But there are other ways you can measure the success of your message beyond potential new customers. Factors like increased name recognition and brand awareness also will speak volumes about your advertising campaign's ability to contribute to profitability. Digital tools have changed the game in terms of measurability and immediacy of results.

If advertising is not part of your current marketing mix, consider it part of your sales efforts going forward. Advertising can be a successful means to building your brand, if you properly develop your message and use it to support a full marketing campaign. By developing a clear call to action, you can drive readers to using your services and/or products.

Jargon Junction

Media Planning Process: Includes goals and objectives, strategy and tactics.

Advertising Plan: An advertising plan identifies campaign goals and objectives and how to determine if a campaign was successful.

Advertorial: An advertisement that has the appearance of an editorial

AIDA: Attention, interest, desire, action. A classic definition of how an advertisement should work.

Assessment/Evaluation: Measurement of media efficiency and effectiveness.

Audience: The targeted group of people you expect to reach with your message.

Buying Media Efficiency: This is a review of media, audiences and rates.

Comparative Advertising: An ad strategy consisting of comparing one brand to a competitive brand.

Cost-Per-Thousand (CPT): Measure of media efficiency. CPT represents the cost of achieving a given reach.

Frequency: How often publications are published. Also refers to the number of times an ad runs in a publication.

Insertion Order: An authorization for a publisher to run a specific advertisement in a particular print publication on a certain date at a specified rate.

Media Kit: A package put out by a publisher promoting their publication. It usually includes a rate card, editorial calendar, BPA statement, ad specifications and sample publications.

Media Strategy: A marketer's action plan to deliver advertising messages to the attention of the targeted audiences through the use of appropriate media.

Medium/Media: Available advertising mediums (publications, web sites, radio, billboards, etc).

Reach: A measure of coverage or penetration of audience. For example, if a publication has a circulation of 70,000 readers, the reach would be 70,000.

Vertical Publications: These are publications whose editorial content focuses on the interests of a specific industry.

Interactive advertising requires additional terminology and metrics to measure performance of marketing activities:

- **CPM (Cost-Per-Thousand) Impression**: advertisers pay for exposure to an audience. Similar to print advertising.
- **CPV (Cost-Per-Visitor)**: advertisers pay for specific visitors delivered to a web site.

- **CPC (Cost-Per-Click) or PPC (Pay-Per-Click)**: advertisers pay each time a user clicks the ad and is redirected to the advertiser's web site. Advertisers don't necessarily pay to have the ad run, but only pay when the ad is clicked. This form of ad is very common on popular search engines, such as Google and Overture. They also run bids for certain keywords – the higher the bid, the higher your site appears on the search results pages.
- **CTR (Click-Thru-Rate)**: the number of click-throughs as a percentage of overall impressions.
- **Conversion Rate**: sales generated from online leads.
- **Affiliate Marketing**: advertiser runs ad campaigns with a large number of publishers and only pays advertising fees when traffic is generated. Affiliate ads are typically distributed through an advertising network such as LinkShare or Google AdSense.

There are a variety of internet advertising options as well:

- **Web Sponsorships**: web site sponsorships can include a space to place the logo and company message, or content sponsorship where the advertiser provides more content and branded ads.
- **Banner Ad**: an ad appears at the top or bottom of a web page. Skyscraper ads are similar, but are vertical in orientation. These can be static or animated ads. Banner ads typically include links to the advertiser's web site or a landing page.
- **Floating Ad**: an ad that moves across the user's screen or floats above the content.
- **Expanding Ad**: an ad which changes size and which may alter the contents of the webpage.
- **Pop-ups, Pop-unders, Pop-downs**: an ad or web page that appears in a new window. These can appear at the top or bottom of a page, and some appear full screen. Research has indicated that users find these intrusive and annoying.

Job Site Branding

Even if your company is diligent about marketing along all the usual channels, you may still have overlooked some aspects of a complete marketing approach. You should seize every opportunity you can to increase awareness of your brand and this includes marketing in ways you might not normally consider. For example, with everything you have going on for a typical project —locating as-builts, coordinating material deliveries and subcontractors, dealing with unexpected changes and budget issues — marketing may be the last thing on your mind. However, project sites are where you should be capitalizing on an oft-overlooked but highly effective aspect of marketing: job site branding.

Job site branding can be a crucial element of project success. Not only will having your company's logo prominently displayed at the job site result in extensive visibility and recognition for your current project, but such exposure also can help secure future work. It might even catch the eye of future employees. Plus, it will give your current team a sense of pride in a job well done. Simply, job site branding is inexpensive billboard space — it's a cost-effective and efficient way to promote your company and gain considerable exposure.

The most obvious way to increase brand visibility on the job site is with signs posted on stationary objects like buildings, fences and trailers, but you can go a step further and place your corporate logo on everything, including pick-ups and ready mix trucks, equipment, tools, gang boxes, and hard hats. Before you start slapping your logo on every flat surface though, it's important to get permission from owners on jobsites. For jobsite signage, this process can start as early as the contract phase, as permission to display signage can be included in contracts. Another step in the job site branding process is to have a variety of signs (different sizes and materials, such as vinyl, wood, and metal) available.

Beyond displaying your company's logo, the way you conduct business at the site will also reflect the image you project, so make sure your team is putting its best foot forward at all times. Maintain a neat, clean site and take pride in your work — both of these will telegraph a clean, professional image that ultimately transfers to the customer. In addition, put a premium on safety and reinforce this commitment with signs around the job site and regular messages to the on-site crew.

Branding employees also contributes to the image of the company. Invest in quality shirts, hats, jackets and other items for employees to wear on the jobsite and on personal time.

Collateral/Sales Tools

While the evolution of technology and new media has had a significant impact on how firms in the AEC industry market, there is still a role for collateral materials such as brochures, sell sheets (typically a front-back flyer summarizing the company and or its products/services), qualification packages and other printed materials. Brochures and/or sell sheets can be a very effective tool for introducing your company, listing your capabilities, products, toolboxes and services, citing references and project case studies. In addition, a well-designed corporate brochure can be a sign of credibility and experience. In many cases, a corporate brochure is still at the core of a firm's marketing communications efforts. In addition to the printed version, a PDF version can be effectively used as well.

There are many instances when having something to send along or leave behind will best assist you in reaching your sales goals. Too often though, a marketing person will identify a need to promote their company and decide a brochure is the answer, though there may be other more suitable mediums. A collateral project should begin with a review of the goals, objectives, message and desired response. If this review validates the need for a printed piece, it is important to then outline the message.

There is wide disagreement about whether the design or the text should come first. It is the opinion of the authors that, once the key messages are identified, the copy should be written, but with an eye toward the design. For example, you should have a good idea about the length and style of the piece, to ensure the copy works well with the graphic tone.

It also helps to engage your audience by planning many entry points into your information. It can help to hold readers' attention by writing copy that has many breakout elements – pull quotes, lists, boxes and more. This keeps the reader from having to wade through each paragraph.

Graphic design can also help you get your message across without being decoration. It can be content-driven and give readers a sense of your tone. It also helps to design with your audience in mind. Design will vary among intended audiences such as men or women, young or more mature.

Wading through the design and production process can be very easy, especially when you work with a professional graphic designer. It is the designer's job to take your ideas and thoughts, translate them into a workable design and give you the results you desire. The process is much smoother, however, if you help the graphic designer to understand your goals and style of communicating with the target audience.

To begin the creative process, you should provide the designer with your finalized copy as well as any art elements such as your logo, photographs and more. Giving the designer a hierarchy of importance for your different elements will also help the designer organize the information. It may take awhile to get a feel for how much you can include in various projects. Telling the designer what is necessary and what may be expendable will help eliminate problems with trying to include too many things.

You also need to communicate your goals of the final piece. This helps the designer to conceptualize a design that will help you meet those goals. In addition, it is important to express any ideas or likes and dislikes you have – it can be as simple as colors you would like to see, layout ideas or anything else that comes to mind. This will help the designer to create a product that is functional, creative and meets your preferences.

Other Collateral

Be sure to share any other design work you have completed. This can help the designer stay consistent with the tone and image you have already put forth to your audience. It is important to stay constant with what you have already presented unless you are changing your company's image and philosophy. This allows your audience to continue to identify you easily. Staying consistent with the tone of your material also helps stay consistent with your company's brand.

Printing

Share with the designer how you intend to print the piece. Printing plans can affect the design greatly and it is helpful to know upfront how the document should be set-up. Decide at the start if you will be printing the piece in-house or on a press because a designer must create and save a document in different ways for each method. Most in-house printers print best when using documents created and saved in an RGB color format while a professional press uses a CMYK color format. Producing and saving the documents in the wrong format can affect the color of the final piece.

While in-house printing is a cost savings, it can also present some limitations in the design. When printing in-house, you cannot use spot or specialized inks. While printing on a press does add an extra cost, using spot or specialized inks allows you to guarantee the color will be exact each time you print it. Many company logos are designed to use spot inks in their original form. Printing in-house or printing on a press without using these spot inks will guarantee that the color of your logo will not match the original color.

In-house printing also limits the graphic designer because it does not allow any element to bleed off the page because it requires a margin on all sides of the document. A bleed is where a photo or color runs right up to the edge of the paper. This is created when the piece is printing on a press and is then trimmed to get rid of any white margins. In-house printers cannot do this because they need the margins in order to grab the paper and push it though.

If you are planning on using an outside printer, consider getting the outside printers involved early in the process as they can usually provide insight that will save money and result in a better product. Consider talking with different printers, online printers and brokers when starting a collateral project. Price should not be the determining factor; there are certain printers that are right for certain jobs.

Deadlines

While the project is in its early stages, communicate your deadline needs to the designer. Many designers juggle multiple projects at once. Knowing your deadline expectations in the beginning will help the designer schedule his or her work accordingly. Once the designer has created a layout, he or she will send you a copy to check for errors. This is your opportunity to express likes and dislikes about the design and it is better to express these now than be unhappy with the finished product. Designers understand that everyone has different visual preferences.

Proofs/Review

As you are checking the piece for typos in the copy, it is important to have more than one set of eyes look over the work. However, it is helpful not to have "too many cooks in the kitchen." When reviewing copy, have people who have not seen the piece before do a review. They are more likely to pick-up errors because they are not anticipating what they will see next. There are several proofing and editing tricks that can help you in this process.

As you can see, there are many factors that go into producing a design that will work best for your company's message and needs. Above all, asking questions and being upfront about your goals and expectations are the best and most important ways to achieve a finished product with which you are happy.

Qualifications

Another important part of the marketing process are qualification packages. While qualification packages are not required in all parts of the industry, they are critical in many parts. Expectations for qualification packages vary in market segment and client. When required, qualification packages can be key to winning projects and new clients. Qualification packages function as a means of establishing your firms' ability to successfully complete a project.

The qualifications are a prime opportunity to make a positive first impression, position your firm and help move forward to the next step in the bidding process, which is typically the Request for Proposal (RFP). Remember, you are making a business case, and that case needs to be supported with relevant facts.

Qualifications have become more sophisticated in terms of the information contained in the proposal and how it's presented. While there is not a rule that applies to all qualifications, it is important that qualifications be professional and client focused. Remember, the client is making a decision, and your qualification package should help them decide why they should select you. More accurately put, they are looking for reasons to disqualify you.

One person should be responsible for assembling all of the materials that will be included in the qualification package. It is essential that the qualification be consistent and flow well. There should be a natural order of items. Content and copy must be clear, concise and to the point. Consider using headlines, sub-heads, and callouts to help readability. It also needs to be free of typographical errors, spelling and grammatical mistakes, and void of marketing fluff. To ensure quality, also use a proofreader.

Many firms have standard qualification packages and "boilerplates" set-up. While this is good from an efficiency perspective, it is important that each qualification was specifically prepared for that specific client and that specific opportunity. Qualifications need to position and differentiate your firm. This starts with researching and understanding what is important to the client.

Graphics and aesthetics are also important when developing qualifications. The graphical look should help differentiate, but must also be consistent with the overall identify of your firm. There should be consistency with your firm's other collateral materials. The qualifications should have an integrated look in terms of fonts, colors, layout, and other visual aspects. Consider working with a graphic designer that can help develop a template for your qualification package. Good quality photographs can also enhance your position. However, make sure that the images used are relevant for that specific client.

While qualifications will vary by client and project, there are some basic requirements that should be included in every qualification:

- Basic firm profile information (contact information, location)
- Scope of services and capabilities
- Number of employees
- Specialization (by discipline)
- Key clients
- Key Completed Projects and Case Studies (Relevant to the client)
- Relevant photography
- Awards

Putting together a qualifications package should not be overly burdensome. Start by establishing a process for qualification preparations. Then develop templates, boilerplates, and content for different applications. Also make sure the staff affords appropriate lead time for qualification preparation. Each qualification package is an opportunity to take a step closer to new business and it should be treated as such.

Jargon Junction:

Collateral Materials: Sales materials, brochures, specification sheets, catalogs, etc., distributed to prospects and customers by a sales person.

Four-Color Process: A printing process that combines differing amounts of each of four colors (red, yellow, blue & black) to provide a full-color printed item.

Pantone Matching System (PMS): A system that precisely characterizes a color, enabling the color to be matched by different printers. PMS colors help maintain consistency in appearance.

Serif Type: Typeface that features short, decorative cross lines or tails at the ends of main strokes in some typefaces, such as Roman lettering.

Sans-serif Type: Typeface of lettering with no serifs, or cross strokes at the end of main strokes.

White Space: Refers to the use of open, unoccupied space of a brochure, advertisement, including between blocks of type, illustrations, headlines, etc.

Public Relations

Many companies make the mistake of thinking public relations involves merely sending out a new employee press release every now and then. While this activity is undoubtedly important, true public relations is so much more — it is a science encompassing a number of varied methods for communicating your message to the public. However, it doesn't have to be complicated if you follow some simple rules for success.

Understanding Public Relations

In its simplest terms, *public relations* is a formal way for organizations to communicate with the public. It is planned or managed communication — a means to communicate, influence and sometimes even sell. It is a medium that provides the third-party credibility you cannot achieve with advertising alone and it is a cost-effective means to build a brand. However, it is important to recognize the other key difference between public relations and advertising: an advertisement is a message you create and purchase while public relations affords you the same opportunity to influence the public, you do not maintain control of your message. Public relations also requires a consistent and concerted effort — it is not as simple as sending one release. True public relations success is achieved by developing an integrated program that carefully delivers your key message to target audiences.

Your Goal

Without a goal, how will you measure success? There are a variety of methods to measure the effectiveness of your PR efforts – ranging from comparing the physical space received with the cost of buying equal advertising space to expensive market research of those reading target publications. For a start-up program, simply identifying key stories and key publications or audiences you would like to target is the reasonable means to measure your success. Your goal should not merely be ink and coverage, rather delivery of your key messages. Think of it this way – your firm does not strive to develop proposals and respond to 100 RFPs in one year. Rather, your goal is to get 25 interviews as a result of your proposal efforts. Develop this same mindset for public relations efforts and set a goal based on the delivery of your firm as a solution.

Assembling Your Plan and Media List

A public relations program requires the same planning effort as any other marketing or sales effort. Start by figuring out what stories you have to tell, and then determine who would have an interest in your message. Use that information to develop targeted media lists of outlets that are likely to give coverage to your stories. In developing these lists, be sure to go beyond the obvious business and daily newspapers and include association newsletters, industry trade magazines or web sites, as well as market-specific pieces. What publications are sitting in your office? More important, what publications are sitting in the office of your current and prospective clients? For example, a general contractor who wanted to brand his company as a solution for precast concrete parking garages not only targeted local business and real estate publications, but also secured placement of his expert piece on how to plan for parking garage growth in healthcare magazines, since many healthcare organizations have a parking shortage. He also targeted and secured publication in a magazine read by parking professionals.

Maintaining media lists is key to any organization. Your media list should contain publication name, complete editorial contact information, web site address, key information on topics found on the editorial calendar, style and editorial theme information, circulation, as well as how editors like to receive information. Don't mail a release to an editor that prefers fax, and don't call an editor who prefers e-mail. Learn their preferences, likes and dislikes, just like you would any other business prospect. And, as you would with any client database, be sure to update this information as often as you can to make sure you are always contacting the correct person.

Developing Your Message

Some of the best stories can be found through conversations with your staff. Was your project the first in your area to use a revolutionary new technique or piece of equipment? Did your team work together to overcome extraordinary obstacles—such as unusual weather, budget or schedule constraints? Can you tie a project or new product to a larger trend in the industry? Will a new building you are involved in constructing have an impact on the local economy? Present your project in light of its newsworthiness, and it will likely catch an editor's attention.

However, just because a project is newsworthy doesn't mean it will be a fit for a particular publication as many are targeted toward specific segments of the industry. To decide whether your story will be a fit for a particular magazine, take a look at a few back issues and get a feel for the type of stories the magazine covers — for instance, are they all about equipment and specs, or do they focus more on the business side of things? It may sound counterintuitive, but targeting your pitch to just one or two niche publications can often bring more success than if you shop it around to as many titles as you can find because editors value pitches that are geared toward their publication's mission.

Once you've decided what you want to highlight about your company and have targeted publications to send information to, the next step is to decide how to package the information. Depending on how much information you have, this can be done in one of several ways: a press release, a press kit or a story pitch.

The most informal of these is the story pitch, which is recommended if you're just beginning a project and want to bring it to an editor's attention for possible future coverage. Story pitches should be as brief as possible; the idea is to give the editor only as much information as is needed to decide whether the story merits coverage. Jot down a few of the project's top selling points and explain each one, clearly and concisely, in three to four sentences. Top this off with an introduction that includes basic information about your company. If possible, include any background information (brochures, pamphlets, previous published articles) — if your pitch is accepted, these will be invaluable resources for the person writing about your company.

Press releases and *press kits* are more formal ways of introducing a story to an editor, and will typically be the format used by professional PR firms. When putting together a press release, remember to approach it differently than you would when crafting an ad or a brochure — the idea is to present the information you have to share in an unbiased, newsworthy format. If you don't have a good writer on staff, don't be afraid to hire one, but remember that publications might not run the information you send them word-for-word. Always date your press releases and include your contact information. (Nothing is worse than sending information to a publication and then being unreachable when an editor has interest in your story.)

If you have a lot of news to present at once — if, you're introducing several new products — or you just want to make an editor aware of your company, a press kit is the best way to do this. A press kit is a folder, CD or flash drive that contains several press releases, plus basic background information on your company such as fact sheets and bios.

No matter how you choose to present your information, (story pitch, press release or press kit), always be sure to include plenty of quality photographs/images as they will make your press release, story pitch or press kit stand out from the pack. If you're sending materials electronically, make sure any attached files are small enough in size that your e-mail won't be bounced back. Some editors may reject large attachments from anyone they don't know, so if you are sending materials to an editor for the first time, you might consider sending images via a ftp site or service such as Yousendit.com.

Working With the Media

According to industry editors, it's often not a lack of good information that causes PR efforts to fail; it's simply the person contacting the editors doesn't possess good people skills. Editors are like any other businessperson — they have deadlines, families, bosses and feelings. Treat them with the same respect you would any other industry colleague or sales prospect. This includes being honest, positive and respectful of deadlines. If you don't have an answer to a question, tell the reporter or editor you will find out. Inquire about their deadline and then make sure you meet it.

It's important to maintain relationships with editors, as this will increase your chances of getting your company featured in their publication. When you have sent a release or story pitch, it's important to follow up, but don't just call to see if they received the release. Call with industry information and details on why the release is a fit for the publication. Persistence is important in getting a message out, but there is a fine-line between persistence and pestering. Don't operate under the guise that the editor will finally give in — chances are you have already burned a bridge at that point. Serving as a resource for an editor is a great way to build the relationship. Most editors and writers today know a little about a lot of things, so educating them is encouraged.

Finally, don't make the mistake of thinking a purchased ad gives you the right to demand editorial coverage. Most publications keep a strict distinction between purchased advertising and the editorial coverage their readers deserve. However, if you do receive an editorial opportunity as a result of your advertising dollars, recognize that this is an opportunity to clearly educate your target audience about your unique selling proposition in a third-party manner. If you are given the opportunity to decide where your ad will be placed, consider putting it on a different page than your editorial coverage so it does not appear as if you "purchased the article."

Industry Awards

Getting your company featured in the media is only one way to build awareness and create a positive impression for your firm — taking advantage of industry awards is another great way to get your name out and build credibility. Nearly every trade association and magazine sponsors an awards program that highlights top construction projects across the country. However, success requires research, preparation and exceptional presentation if you hope to make it to the winner's circle and gain esteem.

Start by identifying potential award-winning projects as early as possible. This will help you plan for photography and other documentation required. You'll also spare yourself a lot of unnecessary deadline stress if you begin well before the submissions are due. Take time to understand the requirements before you dive in. If you are not sure, contact the sponsoring publication or association with any questions you may have.

Once you understand the submission requirements, make sure you follow the rules and include all required information. The requirements and specifications are there for a reason—never assume you'll be making yourself look better by circumventing some of the rules. Also find out what criteria will be used to judge submissions, as that will help you tailor your presentation.

Remember, the more people you have working on an awards submission, the easier it will be to put together. Consider working with a supplier, an engineer, an architect, a contractor or another company that contributed to the project. They can help share the effort and costs, bring additional resources, or even provide a different perspective. And don't forget to include your owner — most award competitions require permission from the company's owner as part of the submission, but take it a step further and involve him or her directly in the process. In most cases, the owner will support and even appreciate your efforts to showcase his or her project. However, be sensitive to any concerns he or she may have. This interaction can help forge a closer relationship.

Gathering all of the relevant information can be challenging, but engaging a team of internal and external co-contributors can improve the process. Consider interviewing a healthy cross-section of those involved in the project. Ask engaging and intriguing questions, and then start to build your story in a clear, concise manner. Be sure to include plenty of engaging photographs that will help tell the story. Depending on the requirements, consider including illustrations, schedules, budget, newspaper articles and other items that help tell the story and set the project apart from others. Consider reviewing award winners from previous competitions to determine what was successful in the past. Also make sure proper credit is given to all parties involved in the project.

Once you've done all that work, don't let your submission fall by the wayside once you've sent it in. Award program submissions can be edited and turned into technical articles, project reports, case studies, direct-mail pieces and web content. If you win the award, be sure to include that in these marketing tools. Even if the award is not a winner, you can still get value by leveraging the content. Once you've done the work, consider submitting it for other award programs. With just a little extra effort, you can customize the submission for other programs.

Client Events

Consider hosting a groundbreaking or ribbon cutting ceremony for projects. These are inexpensive ways of getting exposure for you and your client. Invite the media, politicians, economic development officials and even other clients.

Community Relations

Becoming involved in community events and activities is a tremendous way to give back to the community and increase exposure for your firm. It goes a long way to building your firm's "community brand." From sponsorships to donations to volunteer opportunities, there's no shortage of ways for your firm to become involved in the community local non-profit involvement, charity, business and professional organizations, leadership of groups, sponsorships, volunteer activities and other efforts. In addition, doing pro-bono work or donating services, products or tools is a great way to invest in your community. These activities help build perception and awareness of your firm. The benefits of local branding include: giving back to the community, making employees feel proud about working for your firm, helping generate leads, enhancing the employment brand, and many others

If you're starting a public relations program from scratch, the most important thing to keep in mind is to be patient. Since public relations offers the big reward of third-party credibility in the marketplace, you must first build credibility with the media, which can be a slow process. Once it starts to pay dividends, though, you'll realize all of your hard work was worth it.

Jargon Junction:

Following are all tactics from the media tool kit that can be integrated into a communications plan:

Public Relations (PR): A form of communication management that seeks to make use of publicity and other non-paid forms of promotion and information to influence the feelings, opinions or beliefs about the company, its products or services, or about the value of the product/service or the activities of the organization to buyers, prospects or other stakeholders.

Interviews: A chance for an editor or reporter to interview someone about a particular topic. Powerful opportunities to deliver key messages and gain third-party credibility; however, the danger is the lack of control.

Backgrounder: A brief review of an organization's history, mission, financial support or other information provided to the media with other publicity materials in order to supply basic information that may be used in a news story. A backgrounder is also known as a fact sheet. Backgrounders are especially important to new companies or to companies beginning new PR campaigns. Advantage – with appropriate background information in hand, media professionals are able to make decisions about what news will run in their publications. Downside – A backgrounder alone is not "news" and is therefore not typically published. To be effective, it must be used in conjunction with other PR tools.

Fact Sheet: See Backgrounder.

Discussion Papers: A discussion paper is used to disseminate information or research quickly in order to generate comment and interest. Advantage is that a discussion paper may become the backbone for an editorial in a publication. Downside is that ultimate control of the news story lies in the editor's hands.

Press Release: A news announcement sent to targeted publications with the intent of being published or of generating interest for a news story. Advantage – keeps your organization in the media's eye and is a vehicle for spreading news items to your targeted publications. Downside – used too often, press releases become overlooked by media professionals. Distribute press releases when you truly have news to disseminate. Different types of releases include:

- Action release: A press release designed to call its audience to do something

- Reaction release: A press release distributed in response to an event, issue or topic of concern
- Advisory release: A press release containing advice on a relevant issue or topic
- Photo release: A press release including images; images may be in slide, print or digital format. Always send images with a resolution of 300 dpi (dots per inch).
- Video news release: A press release in video format

Public Service Announcement:

1. Advertising definition: An advertisement or commercial that is carried by an advertising vehicle at no cost as a public service to its readers, viewers or listeners.

2. Sales promotion definition: An announcement aired free of charge that promote either government programs, nonprofit organizations or community service activities.

3. Social marketing definition: A promotional message for a nonprofit organization or for a social cause printed or broadcast at no charge by the media. Advantage – It's a free media tool that creates a positive image for your organization. Downside – It's not the same as an advertisement and may limit your message. A PSA is a subtle, soft-sell approach.

Letter to the Editor: Often overlooked as a media tool, a letter to the editor is an opportunity to voice an opinion on a topic in a magazine or other publication. It is a useful tool for generating awareness about your message. You also can write a letter to the editor about someone else's PR. Advantage – Letters to the editor often have a better chance of getting published than press releases. Downside – Be careful not to write with the tone of an ad, or your letter to the editor will not get published.

Opposite Editorial: Opposite editorial in a newspaper appears opposite the editorial page, hence its name. Op-ed pages generally include both syndicated national columns and guest editorials on timely issues by local experts. These guest editorials are called "op-ed pieces." Advantage – op-ed page is among the most widely read sections of the newspaper. Downside – Because of the newspaper audience, an op-ed topic must appeal to the general public, which may or may not be the most ideal target audience for your message.

Media Contact Sheet: Also called an expert resource list, it serves as a one-stop resource for members of the media when they need a source on a particular topic. The contact sheet contains complete contact information for key interview sources as well as information about their background and the topics for which they are well-suited to speak and comment on. Advantage – With a contact sheet in hand, an editor is more likely to call your organization when in need of a quote or source for a particular topic. Downside – Contact sheets are useful tools for editors, but contact sheets alone do not convey your actual message. Other PR tools must be used to disseminate your message.

Brochure: An internally developed or produced magazine or brochure designed to communicate the views of the organization to a selected audience without outside editorial restraints. Advantage – A brochure is a blank canvas for conveying your message. You have total control of the outcome. Downside – Unlike a published article, a brochure lacks third-party credibility.

Advertisement: Any announcement or persuasive message placed in the mass media in paid or donated time or space by an identified individual, company or organization. Advantage –Total control of your message and guaranteed exposure in the intended media outlet. Downside – Placing ads in media outlets will incur an expense.

Columnist: Some publications will invite experts or authorities in certain fields to write as a guest columnist for the publication. Advantage -- An opportunity to write a column in a publication provides a credible outlet for your message. It is an opportunity to set yourself up as an expert, discussing the latest issues in your field. Downside – A column is not a platform for advertising. To be effective, you must be able to write about a current issue without sounding as if you're pushing your organization's interests, products or services.

Editorial Board Meeting: Serving on the editorial board for a publication can help represent your interests to the media. Advantage -- It is an opportunity to be held as an expert in your field; you build relationships with editors who can help promote your message. Downside – Serving on a publication's editorial board is a time commitment; you must be fully aware of the requirements before committing in order to use an editorial board meeting as an effective tool for your message.

Gimmicks & Non-News Items: Gimmicks and non-news items are often used to create "buzz" about your organization and its message. Advantage – They are especially useful in conjunction with other media tools (at events, press conferences) to increase the impact of your message. Downside – These must be carefully planned and executed so it's your organization and your message that are remembered, not the gimmick.

Media Briefing: Similar to a news conference, a media briefing is a live opportunity to get your message to the media. It may be done in person, on the phone (media call) or via the Internet. A media briefing is useful for shorter, smaller items of news. It may be an update on an ongoing situation or item of news previously announced. Advantage – You have the media as your captive audience. Downside – You have a lack of control over the way your news is presented in the media.

Media Call: A media call is a press conference or media briefing over the phone. At least one expert from your organization speaks via conference call to invited members of the media regarding a news item. A media call may be held in conjunction with a Web-enabled presentation. After the media call, members of the media may be able to ask questions of the expert. Advantage – You have the media as your captive audience. Downside – You have a lack of control over the way your news is presented in the media.

Media Kit: A media kit is a folder of news materials about your organization. Media kits are useful if you have several items of news to disseminate at one time. A media kit may include images, backgrounders, fact sheets, media contact sheets, etc. Advantage – A media kit is helpful when you have more than one item of news. Media kits are generally expected by media professionals if you stage a media event, host a news conference, etc. Downside – Media kits can be expensive to produce, depending on materials used for the kit. An expensive media kit may draw attention to your message or news item, but if the kit doesn't contain real news, it is of no use to the media, and your expensive media kit will end up in the trash. Similar to a press release, a media kit must be used in conjunction with true news items.

News Conference: Members of the media are invited to news conferences where they expect to hear a major announcement from your organization. Typically, members of the media are invited to ask questions of representatives from your organization following the announcement. Advantage – At a press conference, the media is a captive audience for your message. Downside – You don't have control of the way your message is published. In addition, a news conference must be used for real news. Otherwise, members of the media will leave frustrated with wasting their time on the "big announcement."

Research: Your organization may produce research and then provide findings to members of the media for their use in published articles. Advantage – Research sets you apart as expert or authority in your field. Downside – Research alone may be a difficult way to spread your message.

Stage Events: Examples of events include: groundbreaking ceremonies; parties at trade shows for customers and members of the media; a product launch where members of the media learn about the product through media briefings, hands-on demonstrations, etc. Advantage -- Events can generate "buzz" and can capture the attention of the media. Downside – Events can be expensive and time-consuming to plan. An event is only successful if it generates publicity or coverage in the media.

Crisis Communication

One element of a communications plan that is often overlooked is the all-important crisis communication plan. Members of the construction industry are accustomed to having a written plan for a job site accident, but many don't realize that such documents also are essential to guide your firm should the media call you about a crisis or a variety of other scenarios, ranging from layoffs to a contested building project. With a review of the following, you can take the initial steps in assembling the necessary roadmap for communicating in a crisis.

Assemble the Team

Begin the process by collecting a cross-section of people from your organization — accounting, human resources, marketing, project management, etc. You may even consider including your legal counsel either formally as a participant in your meetings or informally through a review meeting once your plan is assembled. Be sure everyone selected for the team recognizes the importance of this committee and its role in helping the organization survive a crisis, no matter how big or small. Obviously, it is crucial that all members are available, and in a position to make decisions and represent their divisions. Don't make the mistake of putting people on the committee based solely on their title or role in the organization. This group should meet frequently so all members are familiar with one another and their styles and preferences should a crisis occur.

Identify the What-if Scenarios

With the proper players assembled, your team should identify the possible scenarios you are likely to encounter. Brainstorm these "what ifs" and then assemble action plans for the scenarios with a detailed list of who to call for what and when. Establish a chain of command, recognizing that it may make sense to alter the day-to-day hierarchy, as successful execution of this plan requires different skills and objectives. This action plan should be reviewed periodically and updated, especially when your firm enters a new market, secures a high-profile project or changes substantially in any fashion. In addition to requiring all members of the team to keep a copy of the most recent plan at their residence, be sure to keep copies of the plan in numerous other off-site locations in case of disaster at your corporate headquarters.

Identify Who Should Talk to the Media

Contrary to popular belief, the highest-ranking official at your company is not necessarily the best person to be your corporate spokesperson. Even if he or she is a polished speaker, giving that responsibility to the highest employee in the organization will leave nobody to clean up any blunders or media mishaps. For example, if your president says something out of turn or disseminates the wrong information, there is nobody to retract or set the record straight. Obviously, there are several exceptions to this rule, and it is often appropriate for a president to comment on, for example, an employee death due to a job site accident. However, such remarks should be confined to simple condolences, while the cause and explanation should be delivered from someone lower in the company in case controversy arises. The ideal spokesperson should have a high enough title and position in the company to command the respect necessary to serve as a credible source, yet still be able to say he or she doesn't know the answer to something.

Decide What You're Willing to Say

By clearly identifying what you can disseminate as a general policy, you will be able to respond to a crisis in a much more effective manner. As a general rule, privately-held companies should be willing and prepared to disclose information on products/services active in the marketplace, items of local interest and facts such as corporate employee figures, as well as policies and practices that are well-established. However, a privately-held company should not be expected to answer questions regarding financial projections, operating results, market share, marketing strategy, products under development, legal matters and upcoming changes. That does not, however, preclude the media from asking about these matters. Document the answers to standard corporate facts to ensure they are readily available.

Announce the Plan to Your Company

Once the plan is assembled, it is important to announce the chain of command and procedures internally. This overlooked step often results in the best plans getting thrown off track simply because the person answering the phone mistakenly provided too much or wrong information to a reporter. As such, make sure your front-line employees are made aware of the policies and procedures, as well as the appropriate contacts, so they aren't caught off-guard by a reporter.

Put the Plan to Good Use

Crisis communication is just one element of a public relations program. While crucial to surviving a crisis or challenge, this same structure (documented contact list, policies regarding data dissemination, identification of spokespeople) can be used as a proactive tactic in sharing positive information about your firm. Avoid the trap of using your communications committee merely for reactive tactics by also seeking input on positive ways to communicate corporate information to the community and the industry. Ideas include highlighting community sponsorship and participation in charity events or other causes, new hires, corporate anniversaries and functions, speaking engagements, new technology and comments on trends. This effort also will serve to build a solid working team to handle crisis scenarios in a manner that is proactive, not just reactive.

Direct Response

You can put tons of time and effort into researching target audiences for your advertising and public relations campaigns, and you can be absolutely diligent about measuring sales leads and ROI, but at the end of the day, there's no way to know without a doubt that your message has reached every single person you intended it to. That's why direct response (the process of communicating directly with prospective customers, whether through mail, e-mail or over the phone) is such an integral part of a successful marketing strategy — it takes out the middleman and gets your message directly into the hands of your target audience. However, if it is to be successful, a direct response campaign takes careful planning, with plenty of attention given to both how your message is packaged and when it is delivered.

Direct Mail

Direct mail can be an effective method for reaching target audiences. Not only will a well-planned, targeted program help build awareness of your products and services and establish an image and style for your company, it will also put you directly in front of key decision-makers and move them to make inquiries about your company. No matter how small or large your budget, you can make a direct mail campaign work for you by keeping in mind a few key tenets of successful programs.

Creative Packaging

Enticing your prospect to open your envelope or self-mailer is often the most difficult job — even if you have a great offer, if your prospects aren't compelled to open the envelope, they'll never find out about it. The exterior of your piece must beg to be opened. The most effective ways to do this are to use interesting colors and graphics, or include a catchy teaser.

Another way to quickly get attention is to use oversized mailers, or even tubes and boxes. While these cost more to produce, the return is usually worth it. Research has shown that people take less than a second to determine if they will open your mail. Unique, oversized mailers have a greater likelihood of getting noticed and opened, especially if you're able to include some sort of sample within the box.

Creative execution is critical to getting your piece opened, but it doesn't stop there. What good does it do to have your piece opened and quickly thrown in the trash? To keep your prospect's attention, you need to involve your prospect. The more you can intrigue your prospect, the more likely it is they will respond.

An Offer They Can't Refuse

The best way to get a response from your direct mail campaign is to include a compelling offer. You should push the target's hot buttons, whether it's with a discount on a product or services, a free premium item or simply additional information. Whatever your offer, be sure to make it something of value to the prospect. By making the offer time-dated, you can often move the prospect to action.

Once you have their attention and interest, don't blow it. Make it easy to respond by including at least two of the following: a business reply card or envelope, a toll-free phone number, a fax number, e-mail address or a web site.

Typical Offers to Increase Response

Ultimately, your marketing challenge is getting the client to say yes to hiring you for a project. An easy way to get there is to offer them something they can say yes to that will help position your firm. Here are some examples of offers that help get the client to say "yes:"

- White paper
- Free guide or handbook
- Article reprints
- Book
- Newsletters
- Premium item
- Free evaluation, inspection, or consultation

- Free educational seminar (must not be disguised as a sales pitch)

Target Your Audience

Many companies spend significant marketing dollars on creative design but overlook the most critical element of a direct mail campaign: the list. Mailing a great piece with a great offer to the wrong prospects will only bring disappointment. Direct mail is most effective when precisely targeting prospects with the highest propensity to purchase your product or service. When selecting your list, target prospects with a profile similar to your current customers. With direct mail, you can speak personally to an individual and tailor your messages to specific audiences. For example, you may not want to deliver the same message to an engineer, a property manager, a school superintendent and a hospital facility manager. By understanding what is important to your prospects, you can communicate with them more effectively.

Timing It Right

When you mail can be as important as *what* you mail. Various industries may be subject to cycles and seasons. By researching the buying habits of your prospective customers, you'll be able to reach your targets when they have a need for your products or services. For example, in the construction industry, weather and geography tend to dictate construction seasons. The construction season in Boston is much shorter than in Miami. There are also other drivers to consider, such as business cycles for your clients. Restoration and construction projects for condominiums in Florida are typically in full swing during the summer, since their peak time is typically the winter when the snowbirds from the Northeast come to town. For condominium projects in the Mid-Atlantic areas, construction activity peaks in the spring and fall.

After you've sent out one mailing, don't rest on your laurels just yet — direct mail programs with multiple waves are the most effective. Repetition and timing are essential to getting your message across to your prospects. Research demonstrates that prospects may not respond until after their second or third exposure to an offer.

Measuring the Results

Before you start any mail project, develop a thorough plan and be sure it makes economic sense. Your ultimate goal is to generate inquiries that will turn into profitable revenue. Start by estimating the response you will need to cover your expenses. Like other functions within your company, marketing and direct mail must generate a return on investment. Determine your expected ROI by dividing your expected incremental revenue by your costs. A few months after the mailing is complete, you will need to complete an analysis to determine if the mailing was successful. The analysis should evaluate how many inquiries were generated, how many leads were qualified, how many proposals were generated and how many sales were made. This information is valuable to determining the success of your program.

Direct mail should be part of a synergistic approach to communicating with your audience. Coordination of direct mail, advertising, trade show efforts, e-mail marketing, public relations and other marketing initiatives will increase awareness and response.

TeleResearch

By now, telemarketing has become so infamous that anyone who owns a phone is familiar with the concept. But what you may not know is that, amidst all of the usual complaints of too-frequent calls and interrupted dinners, telemarketing can still be a viable marketing tool in business-to-business environments, provided that you know how to use it correctly. Start by changing your perspective from telemarketing to teleresearch. Telemarketing is more focused on selling something but teleresearch is more focused on gathering information that can lead to a sale.

First things first: Within the construction industry, telemarketing should never be used to try to sell a product or service. Given the scope of most construction products and services, more thought must go into their purchase beyond a simple phone call. Rather, teleresearch is a way of gathering information to help move a sale along, whether that means identifying decision-makers, qualifying leads, setting appointments, following up after a sale, and conducting surveys.

Teleresearch can be effectively integrated in to AEC marketing programs. Professionals are even less tolerant of telemarketing than consumers. Successful firms structure their teleresearch efforts as more of an inside sales effort. The inside sales effort can be an effective part of a follow-up effort or a direct response campaign.

In an AEC environment, teleresearch can take a number of forms, including:

Identifying Decision Makers: Teleresearch efforts can be used to identify decision-makers and identify attributes about the prospect. Once lists are called, they can be used for direct marketing efforts and by sales people.

Inquiry/Lead Qualification: Marketing efforts such as advertising, direct mail, and trade shows can generate numerous leads. Teleresearch can be an efficient means of qualifying leads in terms of decision-making authority, need, timing and budget. This process further develops the lead so that it can be handed off to a business development person.

Opportunity Qualification: In the construction industry, there are a variety of project information sources, including, but not limited to: *Dodge, CMD, Bid Sheets, Industrial Information Resources*, etc. These services offer a variety of information about projects in various stages of planning, design, and construction. Inside sales efforts can focus on verifying the data and gathering additional information related to a specific product or service.

Customer Surveys: Following a purchase or completion of a project, teleresearch efforts can focus on assessing customer's satisfaction with the product or service. Sending out a follow-up questionnaire can be effective, but a call is more personalized. Trained teleresearchers can dig deeper to uncover true concerns, and even identify opportunities for the next sale.

Customer Reactivation: Inside sales efforts can focus on reactivating old or dormant accounts to identify new business opportunities.

Research: Gathering information about prospective and current customer preferences, experiences, expectations and other information is another effective use of teleresearch. Teleresearch efforts can help you better understand needs and interest to better target prospects and customers.

To be effective, each inside sales initiative must define a specific, definable objective. To help reach the objective, a guideline or a "loose script" should be developed for the callers. The guideline should include a list of frequently asked questions, objections and other information likely to be brought up by the target. Since each call can take many paths, it is critical that the callers be fully trained on the company, product and service. Outside telemarketing firms can also be used for a variety of marketing efforts. When using outside firms, it is worth the investment in training and educating the outsourced callers.

Mailing Lists

Having the right list is crucial to the success of any direct-response campaign. Ultimately, you should have a goal of developing your own in-house list, but you will likely need to augment it with lists from a variety of sources.

List services, or list brokers, are consultants that can help you find the right list sources. List owners are the actual people or companies who own a list. These lists could be circulation lists of publications, attendees of trade events, members of associations or groups or a variety of other sources. List selects are components of information about the contacts on the list, such as demographic, geographic, affiliation or historical information. While select lists often cost more, in many cases, they are well worth the additional cost, as a more refined target list typically results in a greater response rate.

Most services don't actually sell lists; they rent them. They are typically rented for a specific use or number of uses. The rental rate typically specifies the usage parameters. For example, a list rental agreement might be for a specified number of mailings, such as once or three times. Be sure to follow the agreement. List providers typically will seed their list with "fake" contacts to ensure the lists are being used as specified. If you plan to mail to the same list multiple times, you will likely need to rent the list again. Most list owners or brokers will offer discounts for multiple usage, but be sure to negotiate this upfront.

Not all lists are created equal.

Years of research has indicated that direct response is a science as well as an art. Testing different lists is a good practice to follow. Testing and measuring response rates will help determine the best tools for future marketing efforts. The United States Postal Service (USPS) offers a CASS certification process. The CASS process improves the deliverability of mail and assesses levels of compliance with postal regulations. A mail list is submitted to the USPS electronically. The USPS compares the list to their database and updates records and adds mailing information. Mail houses can then be used to print the address on the mailers. The USPS offers bulk mailers that use the CASS system a discount on postage. In addition, CASS certification typically improves deliverability of mail. Most mail houses offer database and mailing consulting services as no charge. For more information on CASS, visit: http://www.usps.com/ncsc/addressservices/certprograms/cass.htm.

E-Mail Marketing

Although e-mail marketing has emerged as an effective tool, the use of lists for e-mail marketing can be problematic. Most people do not want unsolicited e-mail, even if you have a relationship with them. Renting an e-mail list and sending out unsolicited e-mail can be inexpensive and might generate a few responses; however, the risks of being blacklisted and blocked from sending future e-mails are too great. It is crucial to get "permission" from contacts prior to sending them information about your company's products or services. Permission Marketing was coined by marketer and author Seth Godin. In his book, *Permission Marketing*[9], Godin described the process as obtaining permission from customers to send them marketing information. These promotional messages need to be based on interest categories, surveys, opt-in, or someone the company has done business with in the past. Permission can be obtained by using an opt-in tool on your web site, or through asking people to sign up to receive more information about your company at a seminar, trade show or during a meeting.

Once permission is obtained, thought must be put into how and when to send e-mails. Define specifically what the goals are and how frequently e-mail is going to be used. How does the e-mail program augment other marketing communications? Consider timing, types of content and messages. Regular e-newsletters are a popular way to communicate information. If a monthly newsletter seems too frequent, consider bi-monthly or quarterly distribution. Develop your newsletter as an informational and educational tool. Instead of just promoting a sale on a specific product, consider featuring an article about different types of end-uses for that product, for example, how to select the right wood for specific applications alongside the special offer. Remember, every time you send an e-mail marketing message, the recipient is making a choice to open or not open it. They are deciding if your e-mail will be a good investment of their time. Factors such as the message subject line and their personal history with your e-mails will determine if the e-mail is opened or not.

Third-party e-mail vendors such as *Constant Contact* and *Exact Target* offer a variety of robust tools that can help measure the effectiveness of your e-mail marketing efforts. From open rate to bounce rates, to analyzing who clicked what links, the analytical tools offered can provide significant insight into e-mail marketing.

Jargon Junction:

Business Reply Card (BRC): A reply card included in a mail package.

Cost-Per-Inquiry (CPI): A standard measurement used in direct response advertising. CPI represents the cost of getting one person to inquire about your product or service.

Direct Mail: Marketing communications delivered directly to a current or prospective customer through the U.S. Mail.

Direct Marketing: Sending a promotional message directly to prospects and customers, versus via a mass medium. Direct marketing includes direct mail and telemarketing.

Direct Response: Promotions request the prospect to directly respond to the advertiser, by web site, mail, telephone, e-mail, etc.

Inquiries: Responses to a company's direct marketing, advertising or other promotional activities. Used as a measure of the effectiveness of marketing efforts.

Mail House: A mail house is a service that provides mailing, database, and sorting services.

Permission Marketing: A term coined by author Seth Godin that describes the process of getting someone's permission before marketing to them. Permission marketing primarily applies to electronic marketing.

Premium Dimensional Mailer (PDM): A mailer or package that is a non-standard size and shape.

Premium Item: A premium provided to the prospect to induce a response. Examples include a free premium item, a discount or additional information.

Response Rate: Number of responses generated from the direct response program.

Self-mailer: A direct-mail piece that doesn't require an envelope for mailing.

Database Marketing

One form of direct marketing is database marketing, which refers to the utilization of databases of customers or prospective customers to gather and store information about contacts to generate a more personalized method of communication.

Databases typically contain a variety of profile information about prospective and current clients, including profile details, transaction history, affiliations, relationships, preferred project delivery method and any information that tracks previous behavior that might be indicative of future behavior. Information is used in a variety of marketing initiatives such as direct mail, e-mail marketing, telemarketing and sales activities.

However, more robust information such as affiliations, contact history, purchase/transaction history, business information, survey responses and other data can lead to more effective communications. Databases also should include information on multiple contacts at a firm, as well as business relationship amongst firms. For example, an engineer could be linked to a property owner. Once information is collected into a centralized database, a number of tools can be used to develop models of a customer's behavior. This information can then be used to develop plans and strategies for communication type, timing and messaging.

Utilizing a tool to capture, store, manage and use information about a market (prospects and customers) to monitor activity and generate the basis for communication is very effective. Information is gathered in a centralized database and a variety of statistical techniques are used to develop models and profiles to understand customer patterns and behavior. These models are used to select potential prospects and customers based on a variety of attributes.

In addition, data can be appended to database records through third-party sources such as *Dodge, CMD, Reed,* and *Dun & Bradstreet*. Data is acquired through a variety of proactive and reactive activities. Information is typically gathered through sales staffs, dealers and distributors. In the business-to-business (B2B) environment, information contained in databases needs to include individual contact information as well as company information. It is important to collect information about the contact's specific role in the organization, not just their title (decision-maker, influencer, etc). B2B databases are very complex and difficult to maintain. Once established, the marketing database can be linked to a company's customer relationship management system (CRM) and financial systems.

Critical elements of any database marketing program include data integrity and cleansing. The currency and accuracy of data must be rigorously maintained.

Getting started with a customer database is relatively easy for AEC firms. A simple, yet effective database can be built using programs such as Microsoft Access, Filemaker or a CRM system. More sophisticated programs can be developed in more advanced programming environments.

Steps to Effectively Building a Database

- Put someone in charge
- Define the expectation
- Provide a tool to collect the data
- Define the data you want to collect
- Centralize the data and make access easy
- Use the data for business decisions

Lead Management

Congratulations — your last marketing campaign was a huge success, and the leads have been pouring in via mail, phone and the web. But now what? Lead generation is not the end of a marketing campaign, but rather the beginning of a profitable sale. After all, identifying and capturing new business is the foundation of sustained success in sales. A defined and disciplined approach to lead qualification and fulfillment is a critical step to ensuring a successful campaign.

Whether through direct mail, trade shows, advertising or any other means, lead-generation initiatives are just the first step in a cycle that can result in a sale. Once generated, a lead must be carefully managed. Not all leads are necessarily prospects, so a crucial component of the lead management process is establishing criteria that qualify a lead as a prospect. These criteria include factors such as interest, timing, purchasing influence and decision-making abilities. Carefully scripted questions — posed by an in-house salesperson or an outside telemarketing firm — can provide information that will help you determine if the lead is a prospect or a suspect. This also can help you determine what additional information the lead might need and establishes the basis for the next contact.

Once qualified, the leads should be distributed to the appropriate salesperson for follow-up. Assigning a ranking to each lead (descriptors as simple as "hot," "warm" and "cold" are easy to use and remember) will help indicate the level of interest or quality of the lead, which will directly influence this follow-up. An effective lead management program also requires a feedback loop, in which the sales team provides the marketing staff with timely and accurate feedback on follow-up activity and conversion successes. Ideally, these processes should be automated through a customer relationship management (CRM) or sales force automation (SFA) system that feeds into a marketing database. Many of these systems have lead tracking and reporting tools built right into the program. However, paper-based processes can be just as effective.

Regular reports on lead generation, conversion rates and closing rates are essential to the administration of an effective lead management program, and ultimately the overall marketing program. Reports can be produced based on a number of factors, including geography, salesperson, market segment, lead type and many others. These reports can be helpful indicators of sales and marketing effectiveness.

There are a variety of prospecting and informational tools available for AEC firms such as: *Dodge, CMD, Hot Sheet, PEC Reports, MDR Alerts, the Commerce Business Daily* and more.

These are excellent sources of information about activity in the marketplace, but they should not be your firm's only means of tracking projects. For many companies in the AEC industry, it is too late in the process once the project appears in a publication. However, these bid and lead services provide an excellent opportunity to track a firm's activity, projects, and work. They also provide solid contact information which can be used in direct marketing and added to databases. Customer relationship management programs such as Salesforce.com also offer lead tracking processes.

News articles are also an excellent source of information about projects. Newspapers and other publications typically cover a variety of projects in all phases from planning to completion. While keeping up with news is challenging, technology can help streamline the process. Tools such as RSS feeds and Google News Alerts can help track specific information based on key words.

Association Memberships

During high school, we were all encouraged to join a variety of extracurricular activities to build our resumes and make us more well-rounded people. As adults and business owners, joining an association can offer some of these same advantages. There is an association for virtually everything. In the AEC industry alone, there are hundreds of organizations, representing every conceivable facet of the industry. These associations play a variety of roles in the marketplace: setting standards such as building codes and guidelines, providing opportunities for networking among professionals with a common interest, marketing and promoting a specific technology, education, research and even political activities.

Chances are, you've thought about joining one of these associations, or are perhaps already a member. What you may not know, though, is how your association membership can also help grow your business. Your membership can be a valuable marketing tool, provided that you take full advantage of it. There is a distinct difference between just being a member of an association and actually being involved. Sure, simply showing up at the monthly meeting may net you a few benefits, but assuming a leadership role, heading a committee, hosting a meeting, speaking and sponsoring events will increase your visibility and stature and will portray you and your firm as a leader in the industry.

Making Your Membership Worthwhile

The easiest way to get more out of your association membership is to take advantage of the considerable networking opportunities that membership offers. Meetings are a great occasion for speaking with others in your industry. Other members may have implemented a new technology or found a way to solve a problem that has been plaguing your company for years. The friendly camaraderie at association events allows everyone to speak openly and candidly about their business. Being apprised of what others are doing around the country is a valuable asset, as it will allow you to stay on top of industry trends. Many annual or quarterly national meetings include trade shows, where you can learn about the newest products on the market (as well as showcase your own products to a receptive audience). The relationships fostered at these meetings can last long after the final session and provide you with new potential sales leads and colleagues who can act as sounding boards throughout the year.

Industry associations are also a great medium for getting the word out about your company's products and services, since zero effort is required to create a target audience — it's already been pre-selected for you. Most associations have printed or electronic newsletters that feature member profiles or project spotlights. Typically, these newsletters go to members, but also to anyone interested in learning more about the industry. Volunteer to write an article for the newsletter, which will position you as an expert in your field. Another method for garnering attention in a newsletter is through advertising. Associations generally offer members a discounted advertising rate and you are ensured that you are reaching a targeted audience. In addition, membership is almost always a prerequisite to submit to an association's award program, which can open up a wealth of public-relations opportunities.

Marketing to Members

Are there members of your association who could be potential customers? If so, donating your product or service — or even a prize — at the next event will help to build goodwill among all the members, while promoting your company to those who could be potential clients. If you don't have a product to donate, see if any sponsorship opportunities exist for the association's next fundraiser, trade show or scholarship program. To reach members more directly, consider developing a pricing model just for members of your association, in which they can get discounts based on volume purchases.

Affinity marketing programs will give you exclusive marketing access to the association's membership. These programs are a win-win situation for everyone involved — you win by getting more visibility and more potential customers, the association wins by getting additional revenue and providing a benefit to their customers and members win by getting a discount. Plus, by partnering with an association in your marketing efforts, you'll be able to focus your message and will benefit from the added credibility that comes from having that association's "seal of approval." Williams Scotsman, a national provider of office trailers and storage units, developed an affinity program with the Associated Builders and Contractors (ABC), the Associated General Contractors (AGC), and the National Association of Home Builders (NAHB). Each organization represents thousands of general and specialty contracting firms. By establishing an exclusive affinity program, Williams Scotsman was able to partner with each association to gain access and exposure to the organization's membership. In addition, the affinity program allowed them to appear to be "endorsed" by the association. In this example, Williams Scotsman offered a free month rental for each trailer rented for six months or more. The associations received a percentage of the rentals. In the end, the affinity program resulted in a win-win-win situation. Williams Scotsman won by increasing unit rentals and decreasing marketing costs; the associations won by offering value to their members and getting non-dues revenue; and the member won by getting a free month's rent. Before getting involved in an affinity program, though, make sure you're prepared to offer a tangible benefit, and that it is in line with the association's mission. Even when marketing on your own, though, association membership can be a valuable tool in leveraging lists for prospecting databases or other sales activities.

Positioning Yourself as an Expert

Ready to increase your visibility even more? Take the podium. Associations often have presentations that have already been developed and are available for members to present at local or state organizations. The chance to present in front of a local school board about the benefits of your construction method for educational facilities could mean a steady stream of new business. Also, associations need speakers for their own events and conferences. By becoming an active member, you are more likely to be considered for opportunities. If you're not asked to speak, consider offering to host an event at your facility.

Additionally, serving on committees will position you for a leadership position within the organization as a member of the Board of Directors. This provides you with the ability to make decisions that influence both the association and the industry.

Referrals are another common benefit of association membership. If someone in your area needs an expert and called your association headquarters, it is likely that you will be referred. The value of this is immeasurable because the client has already pre-qualified the association as a place to seek an expert opinion and their recommendation of your company is an incredible asset to your business. Further, many associations include a listing of members on their web sites and may even provide a link to your company's site if they have a good relationship with you. Free magazine subscriptions are also common to help keep you abreast of the latest industry trends. Associations are frequent attendees at trade shows, which aids in increasing awareness about your industry. Your organization may also receive a discounted admission rate for members who attend the event.

Joining an industry association is much more than simply something to add to your company website or include in proposals. Active membership affords you a host of opportunities to grow personally and professionally. Similar to your high school activities, it is up to you to get the most out of your membership. Investigate the offerings afforded to you through your association membership and determine how you can get involved on a deeper level. Associations exist to advance the industry, which in turn, increases your business. Partnering with them will help you meet your common goals.

Keep in mind, *not all associations are created equal.* When becoming involved in an association, it is important to have a good understanding of the association's purpose and mission. This will help determine the appropriate type and level of involvement.

Photography

Quality photography is an absolutely essential element of any marketing program. Your company may be at the top of its field, having completed many impressive projects and won several industry awards — but if you're touting all of this on a brochure filled with nothing but line after line of plain text, chances are no one's going to be compelled to pick it up and find out. Making an investment in quality photos of your projects will allow you to reap benefits for a variety of marketing tactics.

Before starting with photography, be sure define the purpose. Are you taking the images to document the construction process? Are you planning to use them in a new corporate brochure? Your purpose will have a significant impact on your approach.

Getting Good Photos

We've all seen examples of amazing photography, whether flipping through a magazine or newspaper, or gazing at images on the walls of a museum or art gallery. But you don't have to be the next Ansel Adams to produce good job site or product photos. Indeed, the proliferation of digital technology has made it possible to acquire excellent high-resolution photos at a very reasonable price. You may already know someone who is a hobby photographer and would be more than willing to take photos for you free of charge.

Before attempting to take your own photographs, make sure your digital camera can take high-resolution photos (300 dpi). Often, cameras are set to take lower resolution photos (72 dpi), which looks great on your computer screen but will look blurry when printed. If your digital camera is five megapixels or higher (most basic digital cameras on the market today average five to seven megapixels), you should be able to get acceptable photos, as long as your camera is set on the highest resolution possible. If you are unsure of how to set your camera to take high-resolution photos, you can always use a 35-millimeter camera and scan the photos in at the proper resolution.

As far as taking actual job site photos, you should make them as engaging as possible and try to tell the story of the project. Shooting a broad mix of views and angles (a wide angle that shows the entire construction site, close-ups on specific equipment or jobs being performed) will ensure you have plenty of choices to create a good mix when using the photos in promotional material — the last thing you want is to have a brochure where all of the photos look exactly the same. If possible, visit a project multiple times to shoot photographs so you can capture it in various stages of completion, which will make for a more robust pictorial. Try to capture various stages from the same vantage point. This will help you tell the story and show progress.

Although people can make a photograph more engaging, hairstyles, clothing, and automobiles tend to stamp an indelible date on the photo, making it harder to use for a longer period of time. If you do photograph people, make sure you capture them in candid situations — nothing looks faker than a posed job site photo.

When taking construction photos, an important aspect is safety. All photographs should depict proper and safe behaviors. People in the images should be depicted wearing proper personal protection equipment. If needed, consult a safety professional before using any photographs in brochures or articles. When possible, have a safety professional with you when taking the photographs.

When taking product photos, one of the most important things to keep in mind is lighting. Nothing can make a product look undesirable quite like fluorescent, too-bright or otherwise bad lighting. If shooting in an interior environment (such as a studio), the goal is to make the light look as natural as possible. Since this can be tricky for amateur photographers, you may want to try to shoot the product outside (though make sure you're working in an area where the sunlight is diffused, rather than a direct glare — early morning is often the best time to get soft, natural sunlight). If possible, instead of trying to arrange a studio shot, take photos of the product being installed, where the lighting conditions may be more natural.

If you are planning to take construction progress photos, be sure to identify specific locations where you will take pictures from each time.

Consider several different perspectives. Also be sure to capture the date and time with the image.

For longer term projects, consider mounting cameras that take pictures automatically at regular intervals. There are many webcam technologies that can effectively be used to document construction. These can easily be viewed remotely through the Internet and posted on collaborative project web sites.

Professional Photographers

If you're unsure about your ability to take quality photos, you can hire a professional photographer. Yes, this can be costly, but the photographs will be top notch — and remember, you'll be able to use them multiple times, for a variety of different marketing efforts. The quality of professional photography easily exceeds what can be gathered by the best amateur photographers. Remember, it is your company's image, so invest in it. Professional photographers will take a variety of factors into consideration such as lighting and angles. When selecting a photographer, ask to see a portfolio of their architectural and construction photography. Before the day of the shoot, visit the site to scout the location and take sample pictures. Also meet with the photographer prior to the shoot to describe your goals and objectives. Discuss how you will be using the images. Show the photographer some of your previous materials. To further cut down the cost, look for opportunities to collaborate with other members of your design and build team to share the photographer's fee.

Using the Photos

Once you have good product and job site photography in hand, the ways in which you can use it are limitless. Obviously, you'll want to use it for any marketing materials you're designing in-house, such as brochures and web sites. You'll also want to send out any relevant photography with press releases so magazines have images they can use with their articles — this will help your story stand out in the eyes of editors, as most magazines do not have the time or the resources to take their own pictures. If you have a large bank of images, even consider setting up a download center in the press section of your web site so media members can select and download the images they need for a particular story.

You must exercise caution when using images, particularly those taken by a professional photographer. When negotiating with professional photographers, make sure you own the images and have full rights for access and use of the images with limitation or additional fees. If you own the images (i.e., if they were taken by a staff member, or if you purchased the rights from the photographer), you may want to impose limitations on their use before you give them to third parties. For example, if you are sending images to the media or are allowing them to download them directly from your site, always provide photo credit information, and consider asking them to sign a terms of use agreement, ensuring the images are only being used to promote your company.

Be sure to keep a thorough archive of all images so they can be accessed at any time by any member of your sales or marketing staff. Consider establishing an index and naming convention so that images can easily be accessed.

Stock photography can also be a viable option. There are a number of on-line sources offering a variety of AEC images. Before purchasing stock images, be sure to review the terms and conditions of use. While stock photography can be a good option, use it sparingly. Keep in mind everyone else has access to the same images, including your competitors, and it is embarrassing to find out someone else, especially your competitor, has the same photo on their materials.

Portrait Photography

Since image is a critical part of you firm's brand, consider commissioning professional portraits for your professional staff. Professional portraits look far better that the digital mug shots taken with the wall as a backdrop. Portraits with a professional backdrop and even portraits at job sites present an image of quality and professionalism to your customers. To get started, hire a photographer for the day to come into your office, or consider setting-up a relationship with a local photography studio. Portrait photography should be part of your new hire process. Provide direction to your staff on what to wear. You might want to consider multiple versions. For example, require suits and ties for men and business suits or dress for the ladies and another set of pictures in business casual attire with the company logo. You should also consider having standards for poses and expressions. Remember, people buy from people. Portraits are a great opportunity to create a positive image. These portraits can be used for proposals, new releases, presentations, web sites and more.

Jargon Junction

Ambient Light: The light that is available. Ambient light is existing light in an indoor or outdoor setting that does not require additional illumination.

Artificial Light: Light that is not from a natural source. Additional lighting is typically from flashes or tungsten lights.

Bit Map: A digital image that is created from columns and rows of pixels.

Composite: The process of combining two or more images. This is typically done with a photo-editing program or graphics software.

CMYK: A color system based on the four colors used in color printing: Cyan, Magenta, Yellow and Black. Can also be a color mode used to define colors in a digital image. CMYK is used when preparing digital images that will be printed using the process colors by a printer or publisher on a four color printing press.

High Res: High resolution refers to the size of the image and quality.

Image resolution: In digital photography, it is the number of pixels displayed per inch of printed length in an image, measured in dots per inch (dpi) or pixels per inch (ppi).

Image Size: Image size refers to the dimensions of a digital image, most clearly expressed in its pixel count, horizontally and vertically.

JPEG (Joint Photographic Experts Group): A file compression method used in digital photography that shrinks a file's storage size. This can also cause image degradation as a result of data loss.

Low Res: Low Resolution refers to a low quality image with a smaller file size. Low-res images are typically good for digital applications only.

Mega Pixel: Refers to the resolution of a camera. A 4 mega pixel camera has a resolution of 4 million pixels.

Meta Data: Refers to electronic information that is embedded in images.

RGB: A color system based on combinations of the primary colors red, green and blue. Cameras, scanners, TVs and PC monitors use RGB.

Pixel: A contraction of picture element. A pixel represents the finest detail that a digital camera can resolve. A mega pixel is one million pixels. In general, the more mega pixels your camera has, the better quality photograph.

Proposals

Gone are the days when a firm handshake or a number scribbled on a napkin can suffice as a project proposal. Beyond building relationships, today's savvy owners require a document that conveys your expertise, experience and professionalism in a mix that is a custom fit for each project. In fact, a recent survey of owners indicates the proposal stage is often when an effective sales process falls apart. Many have found themselves losing work that should rightfully be theirs based on expertise and experience because they fail to assemble a proposal showing prospects the information they desire in a suitable format. However, with some changes in process and organization, plus a little extra attention to the final product, you can increase your chance for success.

Where We Fail

As an industry, we often spend months generating leads but fail to invest much effort during the homestretch process — converting a company from prospect to contract. Owners often cite too much boilerplate information stitched together in an obviously rushed manner, or worse, documents that talk all about you and not about the prospect, as reasons the sales process falls apart. Did you fail to answer a question in the RFP? Does the number of times you use the words "we" or "us" outnumber the times you mention your prospect? Too often proposals have the attitude of a cocky fourth grader with his arm in the air yelling, "Pick me, pick me!"

Beyond submitting information that is too self-centered, owners in the aforementioned study complained proposals often lack relevance to the project at hand. Owners are becoming savvier, and though proposals may have been approved by your key contact in the past, it is common for proposals today to be run past a committee of stakeholders who may or may not know anything about your company and experience, let alone the construction industry.

The key to success, however, is as simple as writing clearly and personalizing the proposal, while paying careful attention to ensure all questions are answered.

Take the recent example of one of the country's biggest concrete firms: Quick to cite their experience on high-profile office buildings in New York City and some of the nation's best-known stadiums, they failed to show any relevant education projects in rural areas when drafting a proposal for a school board. Although they were certainly capable of performing the work, by not responding to the RFP with the information requested, their proposal appeared arrogant at best. The "little guy" ended up getting the job simply because they responded with a custom proposal that clearly displayed their relevant expertise and experience. Even more important — it showed a genuine interest in the owner's project.

Go or No Go

Contrary to the common reaction, the first step in responding to a proposal is not assembling information, but rather deciding whether you should respond at all. Although the Go/No-Go process is frequently overlooked, it is crucial to evaluate each opportunity in terms of your ability to provide a needed service for the client, to provide challenging and rewarding work for your staff, and to make a profit. Other factors to consider include:

- Does the proposed project meet any specific marketing plan goals for profitability?
- Does the project falls in any existing project niches?
- Are there expanded service opportunities?
- Does the project aid geographic expansion or have outstanding public relations value?

Other key areas to review include how much time you devoted to pre-selling, experience with the client, contract issues, similar project experience, availability of your team members, experience with other likely team members, as well as the anticipated fee (including any potential bonuses or penalties). Also be sure to have an understanding about how proposals are being reviewed, the primary selection criteria, the makeup of the selection committee and the anticipated schedule. If you know a "low-ball" competitor will be going up for the same proposal, make sure you can provide a clear non-price advantage in your proposal.

Often, though all logic points to you declining to draft a proposal, the project will offer the promise of future work and repeat business, or has some other intrinsic value that makes the process worthwhile. Sometimes you may even know that the project is a sure thing for a different company, but there is a benefit in getting your information in front of the owner. In such cases, should you decide it is worthy to submit despite the results of the Go/No-Go review, be sure to balance your response effort with the anticipated outcome, and keep the expectations in mind. As the old adage states, you must spend money to make money. However, today's marketing gurus advocate proposal development should never exceed three percent of the potential win.

Getting Started

After making an informed decision regarding whether to respond to the proposal opportunity, begin your response by reading the RFP carefully. Chances are, the prospect wrote it a certain way for a reason, so follow the directions verbatim. If no RFP is available, call for additional requirements. Be wary of those who say, "We just want to see what you come up with," as that same attitude also may be the tone of the eventual contract.

Whenever possible, develop an outline before you start assembling your response. This step will allow you to determine what you need, what you have and where to go for additional information. If you don't already have a marketing database in place, create one to maximize efficiency when collecting project information, references, case studies, photography and corporate information for your proposal efforts. Also, from the onset of proposal development, be sure to discuss and plan the cover letter. Too often, the cover letter is assembled at the last minute and ends up reading more like a transmittal rather than a powerful statement.

The first place to get started is with the questions. Before even answering them, list them word for word as they appear in the request for proposals. Although this may seem silly, most owners prefer to have the questions listed next to the information. Not only does this keep you focused on answering the question, it also keeps the reader focused on the proposal, as they don't have to switch back and forth as they shuffle through papers.

Owners of projects have stated that inconsistency can make proposals seem disjointed at best and unprofessional at worst. For example, if you include years of experience on one resume, be sure to include the same information on the others. In addition to consistency, simplicity is key. Using bullets in lieu of paragraphs is a winning tactic, and be sure to use the word "you" instead of "we" whenever possible. If the number of times you use "we" outnumbers the number of times you mention the client, you lose. Also, be sure to stay away from superlatives. Although you'll win fewer bragging contests, you'll win more clients. Finally, let someone outside your industry read your proposal and explain it to you. This is the best way to test your writing.

What They Read

First and foremost, your prospect looks for reasons to disqualify you and make their selection process easier. Did you follow the RFP? Was your response too long or too short? Was there too much boilerplate information? Although it should go without saying, clear, concise writing is essential. Too often we are so worried about explaining or making our case that we ramble or include extraneous details, failing to answer the questions listed. When responding to questions directly, answer without hype. Avoid the temptation to add detail. The shorter the answer, the better — selection committee members love one-word answers. For example, if asked if your project manager will attend monthly meetings, simply reply "yes." And, by all means, do not refer the reader to another section in your proposal to find an answer.

While you should pay careful attention to every word in your proposal, it is probable the reader will focus on certain areas first — the cover letter, executive summary and pricing. The cover letter should be kept to one page and should clearly discuss the relationship and ask for the work, while the executive summary should concisely provide a recap of the proposal, including the pricing. When it comes to the section on pricing, you'll win points by being straightforward and clearly explaining what is included. Further, don't underestimate the importance of grammar and spelling. Many proposals are reviewed by administrative personnel with a strong English background as the first stage of the elimination process. In this case, even if you have a relationship with someone on the committee, you may be disqualified by someone you have never met because your proposal contains basic errors.

What They See

They say you can't judge a book by its cover, but we all know that your prospective clients will judge you by yours. Like it or not, everyone forms first impressions — it's human nature. If your proposals look like an old college term paper and you're still using the old standard one-inch side margins, stop immediately. The most readable proposals have text running four inches or less across the page with graphics in the side margins or within the text.

Savvy marketers are beginning to pay much more attention to the graphical presentation of their business proposals. Prospective clients facing the task of wading through stacks of proposals filled with thousands of words usually welcome efforts designed to make their lives easier. Graphics can help a reader to better grasp a concept, as well as make your proposal stand out from the rest of the pack. However, the graphics must add meaning and be easily understandable. Make sure your graphics are of professional quality and are placed appropriately throughout the text. People look at images before they look at words, so graphics that are confusing, cliché, meaningless, unreadable or possibly offensive can raise issues of credibility quickly. Before you decide to use graphics in your proposal, check the RFP to see if there are any restrictions or guidelines for their use.

While creativity is applauded in some respects, be wary of using anything but the standard 8.5- by 11-inch size paper or standard binding options, as many odd-sized proposals are hard to file, copy or distribute. Use tabs to separate sections of the proposal, and include the question or heading in bold or italics before your response. And the font also matters. According to *Web Marketing Today* magazine, readers prefer Arial 12 point font (68%) to Times New Roman 12 point font (32%)

From Print to Payoff

With the proper Go/No-Go process, research on the prospect and their needs, as well as sufficient planning, your proposal should clearly show real-world examples of similar experience in an efficient and interesting way. But your proposal also must prove you care about what you do, sending the message you will apply the same enthusiasm and expertise to meet the client's needs. To ensure your message is clear, be sure you have captured the fact that the business is important to your firm. Show respect for the owner, their time and intelligence, as well as a passion for solving their challenge. If you want to be picked for reasons other than price, show the owner that if your proposal is any indication, you care about what you do and you will care about the work you do for them.

Project Reports

How many times have you scheduled a meeting with a potential client or business partner and realized you don't have any materials that show the recent projects you've completed? This all-too-common scenario leads to a scramble to assemble photos and facts, resulting in a presentation that is haphazardly laid out and never proofread before being presented. By taking the time to develop a process for creating project sheets, you help eliminate some of the marketing chaos and ensure your best projects are included in your portfolio.

Getting Started

Project sheets are a simple and cost-effective way to build your portfolio of marketing materials. They demonstrate your expertise, show the diversity of projects you have completed and highlight your experience with high-profile clients or projects. In project reports, content and style are equally important but all information must be presented with the prospective client in mind. The first step in developing a process for project sheets is to begin tracking project data. The tracked data should include information such as full project name, location, square footage, other key team members, estimated project value and completion date. This basic information can be used for a variety of materials, including project experience lists, news releases, project descriptions and proposals. While expensive software programs do exist for this practice, it can also be done simply and effectively in an Excel spreadsheet. Storing the data in one place and having it readily accessible will save countless hours of information gathering.

Once you have compiled the information, review the list and select several key projects that showcase your expertise in a particular area. After you have selected the projects, you can begin to assemble more specific information about each one. Develop a form that project managers can easily complete to garner details about the project. This form should include all of the information included in the project database so the project manager can verify it is correct. Project sheets should include a paragraph — approximately five to 10 sentences — describing each project. Include questions on the project sheet form that will aid in the creation of this paragraph, such as "What is unique about this project?," "What were some of the challenges faced on this project?," "How were you able to solve them?," and "What was the result?" Open-ended questions will illicit more detailed responses. And, as with all other marketing materials, do not underestimate the importance of good grammar, sentence structure, spelling, and clear, concise language.

Looks Count

The final step before you begin creating a project sheet is to ensure you have good photos of the project. Don't underestimate the value of good photography — these images can sell your next project for you. After all the information is gathered, you can develop a template for all project sheets. Consistency is key and a template will ensure you achieve this. Have someone with a good graphic eye create a template that is easy to follow and maintains a consistent look. Many opt to hire a graphic designer to create the template so it corresponds with other marketing materials. Regardless of whether you create it in-house or outsource the initial design, be sure to create something versatile and professional. The template should include your logo, web site address and directions for placing text. Ensure your template is reader-friendly and laid out in a logical manner. Show the template to members of your team and encourage them to provide feedback about the design. Once the design for the template is finalized, stick with it. Resist the urge to alter the design for each project. The template will save a great deal of time and ensure consistency and create branding in your materials.

Key to Success

Although the initial time spent creating project sheets can be extensive, once a template and process are in place, the task should become familiar and easy, and the benefits are numerous. One of the major advantages of developing project sheets is their versatility — they can be used in proposals, as leave-behinds after meetings, as introductory pieces and even for a direct mail campaign. Further, the information you have gathered for the sheets will be very helpful for other marketing tactics such as project profiles on your web site, background information for press releases and the first step of an award submittal. The key to success with project sheets lies in sticking to a defined process for gathering information and creating a sheet for each project in order to maintain an updated portfolio.

Seminars

Maybe it's the fact that more people fear public speaking than death, or that many don't know how to obtain opportunities, but presentations and seminars are an often overlooked marketing tactic. You may wonder why you should bother with trying to fit yet another AEC association meeting into your already hectic schedule. But public speaking is an important part of a marketing program, as it as it not only creates awareness of your company, but also increases your credibility as an expert in the eyes of your target audiences. Building this long-term credibility is essential in maintaining your position in the marketplace, even during slow economic times.

Develop Your Presentation

The first step in capitalizing on this marketing opportunity is identifying topics on which you feel comfortable speaking. A successful topic will be one you have unique expertise to speak about, but also is timely and addresses market trends. Further, it is important to identify topics you can back up with relevant case study examples and practical tips. Not only will these examples bring your viewpoints to life and further add to your credibility, but they also will help you build rapport with your audience.

The cost effective seminars are not promotional events. In order to build third-party credibility, your message should be viewed as market expertise, not a commercial. Your goal should be to inform and educate. Having a strong sales pitch will have a negative impact on the audience and leave a bad impression, turning away potential clients. Understanding your audience and their expectations is also critical to delivering an effective message. Make sure your topic is appropriate for the audience, then further tailor it in terms of presentation style, communication and level of technical information. For example, a presentation to structural engineers might require a higher degree of technical information than one delivered to property managers. Also avoid cramming too much information into your presentation. Rank and prioritize the topics and points you want to cover and sacrifice the information of less importance.

Always be Prepared

With your topic identified and material developed, it is important to properly prepare for any speaking opportunity. There is a huge misconception that too much practice results in a canned presentation. However, studies show practice actually results in spontaneity, because you are more comfortable with the information. As such, be sure to practice your presentation with a wide variety of audiences before speaking to potential clients. Remember not to read the slides, but use them as a guide. Slides should contain bullet points, not a narrative description. Structure your presentation by telling your audience what you are going to tell them, then tell them, then tell them what you've told them. Audience participation can be engaging and create enthusiasm, so let the audience know when you will take questions. Be sure to anticipate questions as well as develop and practice responses. Also consider raising questions during your presentation to provoke thought.

Ensuring Seminar Success

While sales presentations are typically held at a prospect's office, seminars are held at more neutral sites such as hotel meeting rooms and conference centers. Although this venue obviously takes a greater level of planning, it offers a lower pressure environment for your audience. The success of your presentation will depend on getting the right people to attend. To this end, there are many factors to consider. What are the attendees' expectations? Will people from more than one company attend? If so, is there a potential for issues with competitors all being in the same room? If so, consider inviting different disciplines, such as one architect, engineer, contractor, developer and a variety of specialty subcontractors so none of the attendees are competitors. Such an attendee list also creates a unique opportunity for all participants to share in the session, as it creates an open forum for discussion. A smaller multi-discipline group also provides a unique opportunity for you to have one-on-one discussions with prospects. Your goals, objectives and message must be considered when determining your target invite list. Seminar length can vary depending on the format and content of the seminar, but they typically range anywhere from 90 minutes to four hours. Breaks should be scheduled every 90 to 120 minutes.

Although it is important to not pitch your company and its services, be sure to capitalize on the seminar as a chance to deliver your key messages. At the end of the session, it is appropriate to share a short overview of your services, provided it doesn't come off too strong. Be sure to ask for feedback and offer additional resources and handouts to attendees. Also offer to post information on your web site, which is an excellent way to drive audience members to your site for more information about your company. Another way to unobtrusively promote your company is to invite other members of your firm to join you for lunch or a closing session to encourage networking and sharing of ideas. After the seminar ends, be sure to follow up with attendees within a week with a continuing education certificate as well as additional information. Such efforts further position you as an expert and help build your brand.

If you don't have the time or resources to organize an entire seminar, you can build the same credibility by serving as a panel member or moderator for a seminar put on by a trade group or industry association. By sharing the podium, you'll also deliver more value to attendees, as information and expertise comes from multiple sources. Partnering with an outside organization also reduces your planning burden, can significantly add to your credibility and gives you an audience with a defined interest. Be sure to check with the associations you belong to in order to determine if opportunities exist.

Offer Continuing Education Units

Another huge opportunity is to become an approved registered provider through the American Institute of Architects (AIA), Construction Specifications Institute (CSI) or other engineering, contractor or owners associations. In order to become a registered provider, your presentation will need to be assembled in a way that is non-commercial. However, there is great validation about you and your company if you earn this designation and are able to offer the continuing education units associated with these programs. Check these organizations' websites for details on how to apply. Remember, offering education and knowledge is a great marketing tool.

Trade Shows

In any industry, trade shows — events where people from a particular industry can gather to learn about new products, services and share information — are an important forum for doing business, and this is especially true in the AEC industry. In effect, trade shows bring business to life — they allow consumers to see your products and services in person and they give you the opportunity to forge deeper connections with potential customers through face-to-face interaction. In addition, the communal nature of trade shows will also give a glimpse into how your competitors are doing business and can provide some insight into trends within the market at large.

However, trade shows represent a considerable marketing investment. The cost of exhibiting at a trade show can include everything from the price of exhibit space to materials for your display, carpet rental, electric, furniture, set-up and labor, shipping costs and extra promotional materials. If you play your cards right, though, it's an investment that will pay rich dividends. Getting the most for your trade-show dollar involves a carefully orchestrated approach that begins well before the show and keeps going long after.

Before the Show

At least part of the formula for trade show success lies in selecting the best shows to attend. If you spend all of your resources on shows that only a small portion of your target audience will attend, your marketing dollars won't stretch nearly as far.

There are literally hundreds of trade shows in the AEC marketplace. With pressure on ROI, selecting the right shows is becoming even more important. Remember, the goal is to generate quality leads that turn into business opportunities. While there are a number of factors to consider when selecting which trade shows to attend, you must balance the number and type of attendees, show hours, show location, size, promotional opportunities, competitive presence, industry trend and other relevant factors.

Once you've determined the best trade shows to attend, set specific, measurable goals for each show. Goals can be in terms of number of contacts, qualified leads, presentations and other specific metrics. Further, goals can also include vendor meetings, client entertainment, press meetings, competitive reviews or recruiting. Make sure every member of your staff understands the objectives for the show. While educating your staff, provide profiles of anticipated attendees, identify their reasons for attending and identify potential questions and answers. They also need to be able to effectively answer questions, qualify leads and operate lead-retrieval systems.

Be sure to get the word out to potential customers that you'll be at the show and provide them with your booth number. Most trade show production companies and trade associations provide pre-registered attendee lists for free or rental. A few weeks before the show, send the attendees a postcard or letter inviting them to visit your booth. Include an offer for a free item or announce a contest. You should also be able to obtain a list of media and editors who will be attending the show. Consider inviting them to a press conference (if you have news to share), luncheon or other event sponsored by your company to introduce them to your new products. These are great ways to get a more captive and focused audience. If this is out of the reach of your budget, a simple booth tour, in which you meet with journalists one-on-one and talk to them about your company, can be equally — or sometimes even more — effective. Try to send out invitations to the press as early as possible, as appointment books tend to fill up quickly at shows. Many shows also offer promotional opportunities including program advertising, promotional banners and signage, event/meal sponsorship and other items.

Spend some time and money to develop a display that is inviting and engaging — after all, it will be the visual representation of your company's image. Display graphics should quickly and effectively communicate your company name and the services and products you offer. Technology can also be effective for attracting visitors. Trade show attendees make quick decisions while walking down crowded show aisles. Your goal is to get their attention and draw them into your booth. A professionally designed exhibit with attention-grabbing graphics, product samples, video monitors, demonstrations and engaging booth staff are essential.

Depending on the nature of the show, there may be speaking opportunities. The programs associated with trade shows are typically set many months, and even years in advance. Contact show management to explore potential speaking opportunities.

During the Show

At the show itself, you have only a few days to make as many worthwhile contacts as you can, so make the most of it. When people stop by your booth, avoid asking, "Can I help you?" Ask your prospects effective, open-ended questions that will help qualify and quantify their interest. Engage them in conversation. If the prospect is a qualified lead, set the stage for the next step in the sales process.

Many shows offer electronic lead devices which scan attendees' card and collect their information. In addition to this information, ensure that the booth staff collects information about the prospect's interest, timing, next point of contact or any other relevant information.

Trade shows are excellent networking events. If a visitor to your display is not the decision-maker in his company, find out who is. Talk to fellow exhibitors, show personnel, the press and others to identify business opportunities.

Your instinct may be to stay close to your booth the entire time to maximize sales potential, but by doing so, you'll actually miss out on other potentially helpful information. Instead, take some time to walk around and check out your competitors' displays. You can learn a lot by observing what and how they are exhibiting, and how they approach their prospects.

After the Show

Most prospects will be so inundated with literature and brochures at the show that the majority of it will end up in the hotel or convention center trash can. Instead of loading prospects down with literature, collect their information, qualify his or her need and promise to send information within a few days. Once the show is over, make sure you send the information within 48 hours. However, you should still keep some literature on hand for the prospects who want or need to see it at the show.

In addition to sending literature, you should also use telemarketing to follow up on the qualified leads collected by your staff and make the next step toward closing a sale. You should also enter inquiries and qualified leads into a database for future use.

Before you plan to return to a show for another year, you need to determine whether it was a success —and simply counting the number of leads received won't give you an accurate assessment. Take it a step further and count qualified leads and find out how many leads turned into quotes and orders. This process may take a few months, but it will be a valuable tool when planning your marketing budget for the next year. This simple approach, with equal attention paid before, during and after a show, will go a long way toward ensuring all of your future shows are a success.

Web 2.0, E-Marketing and Social Media

A web site is ultimately a collection of data, information and digital assets such as video and images. However, strategy, planning and marketing have become critical factors for web success. As the Internet has evolved, so has the importance of the company web site as a piece of the integrated marketing communications puzzle. However, it is just a piece of the puzzle. An AEC firm's web strategy must include a strategic approach to search engine marketing, social networking and public relations. Today, AEC firms must leverage the trend of using the Internet to communicate and collaborate to achieve its marketing objectives.

In the mid- to late-1990s, it became clear the Internet was here to stay and every company that wanted to survive in the business world needed to have a presence on it. This led to many companies scrambling to throw up a web site just so they could have one. However, the web was usually low on the priority list for most companies, so these sites were very basic and rarely updated. When corporate web sites first started to appear, most were simply online versions of brochures. Then, in the late 1990s, e-commerce increased in popularity as many sites offered transactions. Since 2000, commercial web sites have continued to evolve.

Now that almost every company has a website, there is a new generation of potential clients who grew up on the Internet and the first place they go to learn about a company is their web site. Not only that, but visitors often form impressions on companies based on their web presentation, which is why it's important to have a web presence that is well-designed, user-friendly and on the cutting edge of the latest Internet trends. With a thorough understanding of what should be included on your site and an understanding and daily interaction with online social media, you'll be fully prepared to market your company in the 21st century.

Web Site Basics

The best way to determine if your web site is working for you is to view it as if you are a web surfer happening upon it for the first time.

Does the site invite you to check out the latest news items, but the last news item is from two years ago? Are the photos and design dated? Is it hard to navigate the site to learn more about the firm? If the answer to any of these is yes, it is time to update your site.

Obviously, the content and design of a web site will vary from firm to firm, but there are certain variables that should be constant for every site. A recent study revealed contact information is the single most important content requirement for a web site. This finding indicates basic contact information – including address, email, and phone number – should be placed on every page. Make your site even more user-friendly by including specific contact information on related pages. For example, if your site has a press room, include the names, phone numbers and e-mail addresses of your marketing team along with your latest press releases and photography. If you have a sales section, include contact information for your sales team. The easier it is for potential clients to get in touch with you, the more likely they will be to do it.

Also take a look at the images on the site. Photos should represent the best your company has to offer, so make sure they are clear and crisp. Finding the right size for web photos can be tricky, especially on your home page — large photos can take too long to load, while photos that are too small may be unreadable. Strive for an equal balance between the two. Images are the first things people will see when they visit your site, so you want to make sure that the photos are representative of your firm. In the AEC industry, the use of videos, time-lapse photography and animations are an extremely effective tool.

There are also some general guidelines you should follow regarding content and design. A common mistake many companies make in their content is focusing only on their firm and never addressing the visitor. Your web site should constantly demonstrate how you are able to meet the visitor's needs. It is always wise to include case studies and project reports that highlight how you were able to meet a client's needs. Another way to showcase your expertise is by including client testimonials. We all love to read what other people have to say and satisfied clients are often the best way to sell your services and products.

However, keep in mind when it comes to content, sometimes less is more. We have all visited sites with so much copy they require you to scroll for what seems like an eternity. When developing the content for your site, keep in mind what it will look like on a computer screen. You have a precious few seconds to grab the visitor's attention so make sure your most important points are conveyed on your home page, then direct visitors to sub-pages for more information. The use of pictures and videos will also grab and keep the attentions of visitors. Keep in mind that people who visit websites often look at them in the shape of an F. They look to see the navigation on the top then go down the left side of the page. Then, their eye will go from the middle to left of the page.

While specific sections will vary based on each firm, common pages include about us, products or services, projects or case studies, news and contact us. Keep the copy easy to read and full of information, not just fluff. And make sure you are updating the site frequently. If your web site has a news section, it is important to update the information on a quarterly basis, at the least, although ideally news items should be added as soon as they are made public. Rotate projects and case studies frequently, adding and subtracting as new and better ones are completed. The projects featured on your site should represent the markets you want to be in and showcase the quality of work you are capable of performing. Ultimately, you want to keep the site fresh and give visitors a reason to come back.

To get ideas for the best design for your site, spend some time visiting other sites, both in the construction industry and outside of it. Make a list of features you like and don't like. And remember user-friendliness is paramount in web design. Graphically oriented sites using Macromedia Flash and similar programs can be very compelling, but make sure there is an option to skip directly over a graphic introduction and into your site. Also consider creating two versions of the site (one in Flash and one in basic HTML) so users with slow Internet connections or without the necessary plug-ins can still view your site. Beyond these graphic elements, give plenty of thought to site navigation. Visitors should be able to browse through your site quickly and efficiently. Information should be easy to find, or visitors will simply leave your site and move on to the next one. Be sure to make it easy for users to contact you. People want information quickly and easily. Provide a variety of accessible methods for visitors to contact you, including phone numbers, e-mail and forms on the website to be sent to you.

Getting Visitors to Your Site

Once you've got a great site, it's time to focus on the all-important next step: driving visitors to it. After all, your hard work won't make much difference if no one actually sees it. These days, increasing web traffic is all about search engine optimization (SEO) and search engine marketing (SEM), which are approaches designed to increase the ranking of your web site on common search engines like Google, Bing, Yahoo and MSN.

Search engines have become essential tools for anyone buying or specifying products or services. In many cases, this is now the first step in the information search process. Therefore, higher rankings typically result in higher traffic. Countless studies have show that users rarely go beyond the second or first links and almost never go to the second or third page.

The first step to increasing your optimization is to get an idea of where your company is currently ranked. To do this, plug your company, product name or keywords that people might use to find you into a few different search engines and see where your web site pops up on each one. You might also want to search for your competition to get an idea of where they rank, too. Ultimately, your goal is to place on the first page.

If you're a little lower on the list, don't panic — there are several easy steps that can be taken to improve your SEO. You may need to hire a third-party to do effective SEO for you. But if you decide to do it on your own, the first step will be to include keywords in your title tag, which are the words that appear at the top of your browser when you open a web page. The content in the title tag usually appears in the clickable link on the search engine. This is typically a key factor in search engine algorithms and ultimately the resulting ranking. Strongly consider using synonyms or alternative phrases to enhance search engine visibility. Just by updating the title tag with your company or product name, you can usually generate quick improvements in the results. Blogs, RSS feeds, social networks, Twitter posts and other Web 2.0 applications are becoming increasingly relevant in search engine rankings.

Next, take a look at how each page is presented and sequenced, and make sure it flows in a logical manner. The text on your home page should include good descriptive terms about your product or service as well as common search terms. This will help search engines index your product or service. Instead, the keywords should reinforce themes already presented on your main page. In addition, when you load images and links, make sure these are also tagged with descriptive keywords.

To help figure out the right keywords to use, start by putting yourself in the shoes of a prospective customer. How would they describe what they are looking for? People searching for information relating to a purchase typically use longer, more specific phrases when searching for a product or company. Keywords should include product names, brands, applications, descriptions of common problems and solutions, synonyms and alternative phrases. And remember specific terms will likely yield better results — choosing a general word such as "concrete" or phrases such as "ready mixed concrete" can return hundreds of pages of competing results. However, "polymer modified concrete repair material," "BSL 3 lab engineering," or "hydrodemolition of a parking structures" return fewer results, they are more relevant results. Google's AdWords program offers useful keyword matching and keyword generation tools if you would like to add to your SEO efforts.

Beyond keywords, the overall content of your site also is critical to maximizing your SEO. It is important for your site to have unique content that clearly demonstrates what you do. Outfitting your site with the various sections previously mentioned (news stories, case studies, information about the firm) will provide a wealth of information for search engines to grasp. Include links to various relevant sites such as associations, trade organizations, publications and other companies. This will help enhance the relevancy of your site in the eyes of the search engines. You should also work to post links to your site on relevant sites where potential visitors might go.

A content management system (CMS) is a software tool used to develop, create, manage, edit and publish web content in a systematic and organized manner. CMSs offer ease of managing web content without web programming skills. Many CMSs offer simple tools and icons (typically similar to Microsoft Office) for users to create and manipulate web content. CMSs can be used to manage single or multiple web sites. In addition, multiple users can manage web content based on their levels of permission. CMSs also allow for workflow process management, content approvals, and easy collaboration. Most web development firms can develop a custom CMS and there are a variety of CMS tools available on the Internet for low cost or even for free.

Once you've made changes to your web site, it must be submitted to the search engines. This will allow the search engines to crawl your site, indexing and cataloguing as they go through it. Web crawlers or spiders are programs or automated scripts utilized by search engines to browse the web, create a copy of web content and index information and data that appears in search engine results. The ultimate ranking algorithms include a variety of factors including keywords, word location and frequency, relevancy, link analysis and click-through measurement. What this means is the SEO process is not a one-time event. On a regular basis, search engines change their algorithms. This requires a constant focus on SEO, involving regular changes and modifications, so you may want to consider engaging a SEO or SEM expert to help guide you through the process.

As much work as you put in tailoring your site to increase SEO, there are still certain variables of the process that will be out of your hands. To gain more control over your search-engine visibility, consider participating in paid searches — essentially ads appearing next to natural search results. Paid searches can be useful to help you cut through the clutter of search engine results. In addition, paid placements ensure your ad or message will be seen by relevant audiences. Further, paid searches can be a very efficient, yet powerful way to reach your audience. Ultimately, you will be bidding on keywords. Your bid per click will help determine your ranking on the site. Eventually, your ad's click-through rate and performance will determine position. Google's AdWords and services such as Yahoo Search Marketing and Overture offer easy-to-use tools to help you develop paid search programs, including ad development, keyword recommendations and budgeting assistance. Google is the most popular search engine today with 70 percent of internet users searching through Google. If you decide to use paid search, go with Google before any other search engines as it reaches the most people.

All of these efforts need to be measured eventually. Google Analytics is a free and user-friendly tool that allows you to measure the stickiness of your website, how long people are staying on pages, where they are being directed from and most popular links on the page.

The Future of the Web

The presence of keyword-based advertising on the web (as opposed to the banner ads that once dominated web advertising) is one example of the evolution the Internet has gone through in the past few years. The real star of the next generation of web services (often referred to as Web 2.0) is the idea of user-generated content in the form of social networking platforms, wikis and other innovations designed to facilitate creativity, conversation, collaboration and sharing between web surfers. Although these terms may seem like nothing more than jargon, the evolution of these tools and services offers unique opportunities for firms in the construction industry.

In effect, Web 2.0 turns the traditional marketing formula on its head, opening up even more avenues for promoting products and services. Instead of the traditional company-to-consumer method of marketing used for brochures, trade shows and the like, Web 2.0 enables companies to harness the power of the Internet to connect people to collaborate, share ideas and provide a system of check and balances. The new social media represents a fundamental change in the AEC marketing philosophy, as marketers focus more on facilitating conversations and move away from pushing messages out. Successful marketers will adapt their market strategy and tactics to the way the Internet works today by engaging customers. It is also recommended that companies place an article about their company on Wikipedia. This not only improves SEO, but allows users to learn more about your company. Keep in mind, when writing an article, always remain neutral (it can't sound like a brochure), include many references and publish on Wikipedia from your home computer (Wikipedia may delete an article published from an office computer).

Specific examples of new or improved technologies on the web include social and professional networking, online videos, social bookmarking, RSS feeds or syndication, weblogs (blogs), wikis, podcasts and other methods of enhancing communication. These advancements allow users of the Internet to do more than just retrieve information — they can now participate and contribute to the information. Thus, instead of controlling communication, marketers merely facilitate it. Also, communicate in a conversational manner. With an integrated Web 2.0 effort, marketers can create, edit, shape and influence messages. To understand how your marketing program can benefit from Web 2.0, it's important to understand some of the terminology that relates to this phenomenon.

Social Networking: Social networking sites have exploded over the last few years. A social network consists of individuals tied together based on interests, affiliations, values or other relationships. These web-based forums provide opportunities for communication, sharing, networking, recruiting and even building your business. Sites such as MySpace, Facebook, LinkedIn, Google Wave, Digg, Technorati, Wikipedia and Twitter have become extremely popular with all generations. While these have broad appeal, networks or communities within these sites are starting to consist of people that share common business interests. There are also business-minded networking sites, such as LinkedIn offering the same features in a professional context. An offshoot of social networking is social bookmarking, in which members of sites such as Digg and Del.icio.us can store and share content and links to various sites with other members of their community. Another popular tool is Twitter, which is a micro blog utility that allows people to keep up-to-date on what others are doing. While based in consumer uses, Twitter has tremendous potential as a business-to-business application. In consumer marketing, these social networking tools are increasingly becoming part of the marketing landscape. These technologies are also starting to permeate B2B.

Syndication: The concept of accessing up-to-the-minute content from multiple sites in one place has also become popular. Real Simple Syndication (RSS) allows web sites to syndicate news and regularly updated content such as news and push it out to third-party sites called RSS readers or aggregators. When it is distributed, it is called a feed. The most user-friendly RSS reader is Google Reader. Users can subscribe to a site's feed by entering the link feed into their RSS reader. Consider looking for ways to syndicate anything that might be useful, interesting or informative, such as newsletters, project reports, technical articles, project updates, announcements, and technology or product news. Syndication also contributes significantly to overall SEO strategy.

Blogs: A blog, short for weblog, is essentially an online journal. Sites like Blogger, Blogspot and WordPress have made it free and easy for virtually anyone to start a blog. Bloggers typically provide information through text, but blogs can also support images, videos and web links. A key feature of blogs is the ability of a reader to post a comment. While blogs have been around since the 1990s, their popularity has increased significantly in recent years and more search engines are returning blog posts. To search blogs, use Google blogsearch. For the construction industry, blog topics can include information about products, procedures, equipment, safety or any relevant topic. The *Engineering News Record* site (ENR.com) has added blogs dealing with topics such as estimating, construction, engineering, safety, career paths and insurance. Many construction-related associations and publications are now also offering blogs. Consider becoming a regular contributor to these blogs or start your own. A blog can provide a way to push new content to your site between regular updates, as well as provide a more informal forum for people to get to know your company. Your blog might serve as a means distribute press releases, or as a medium for current employees to talk about your company so it can be used as a recruiting tool. It is also an excellent forum for dispelling information about current projects. For example, creating a blog for a school project can be a place to update the community about the progress and highlight important project milestones. Keep in mind that, though your blog may be more informal in tone, it should still support your brand and marketing messages. Before you take your blog live, try keeping an internal, private blog for a while just to make sure you have proper information to share and will be able to commit to the frequent updates. This will help educate your entire team about what is proper information to share and help create guidelines for an external blog. Also, because blogging is user-generated, there is always the danger that unhappy customers or others in the industry can propagate negative feedback about your company, product or service via your blog or their own. Therefore, you should consistently monitor blogs and comment posts — several companies even offer blog-monitoring services if you are too busy to keep track of it on your own. Lastly, make sure to regularly update content. If you only publish weekly, make sure to publish at the same day and time so users know when to expect new content. Also, be aware of Technorati and Digg as those outlets are used by bloggers to find and share content.

Wikis: The term "wiki" refers to software or a web page that allows users to create, edit and link content. Essentially, wikis allow for content to be written collaboratively and usually require a username and password to access. Many companies are using wikis for collaboration and knowledge management. Consider setting up a wiki inside your company and allow employees and teams to collaborate, post information, pictures, videos and any additional information about your projects. The most popular example is Wikipedia, essentially an online encyclopedia containing millions of entries in dozens of languages, written by users that have knowledge of a particular subject. Wikipedia significantly improves SEO as it is usually one of the first results. It is important to take a proactive approach by adding non-commercial, relevant content to entries on sites such as Wikipedia. Be sure to include references to projects, applications, technologies and other insight or expertise you have on a topic and monitor what is being posted.

Podcasts and Online Video: Within Web 2.0, users are also sharing information via online broadcasts (called podcasts because the programs are downloadable to iPods and other mp3 players). Podcasts are very similar to radio interviews or conversations except they are recorded and downloadable so you can listen to it on the go. Some companies are creating podcasts about construction topics that users can listen to on the job site. YouTube is an online video source which is extremely popular and contains a video for every subject and topic. It allows users to post, view and comment on other videos. The good thing about these developments is that they are free and easy to use. Topic ideas include information about your company, commercials, interviews with employees or top management, tours of offices or job sites, product applications, new product introductions, project reports, safety philosophy, and "how-to" pieces. The best way to gain an audience for your podcast or online video is to make it a structured program with a regular schedule and format.

Leveraging Web 2.0 Tools

With a greater knowledge of Web 2.0 opportunities for today's construction arena, the next step in learning how to use it for marketing is to understand different rules that apply to this medium. For example, community users typically do not respond to traditional marketing or advertising messages. Strong editorial content promoting a product or service is typically highly valued, where as overt advertisements are rejected. A premium is also placed on authenticity. Social marketing is a form of viral marketing, in which interesting things spread more quickly. As such, developing and implementing a Web 2.0 strategy takes diligence, commitment and effort. The content needs to be consistently created and managed. The worst thing a company can do is create a page, group or blog and neglect it. Also, according to a recent study completed by Social Media Magic, companies that are successful with online social media invest 32 hours per month on one profile. It takes one year to see complete success, according to the study. One key way to learn the rules is simply to participate in the social networking Web 2.0 offers so you can gain a better appreciation for how it works and what is accepted in this space. Start by creating a strategy, establish your presence, expand your reach, nurture relations and properly maintain your online presence. Before you start, do an audit, see where your target audience is communicating, then dive in and join in the conversation with them. Also, keep in mind that people join online communities because they are interested in the content. Don't spam and don't use words that make it seem like an advertisement. Treat them like real people who care about things that your company can provide for them.

As with all other marketing tactics, it is essential to create a plan that ties to your overall marketing strategy. But with Web 2.0, different metrics apply. While traditional marketing metrics include circulation, reach, response rates, hits and click-throughs, in Web 2.0, metrics and measurement focus on the number of blog readers, comment posts, search-engine rankings, references, contributions to blogs and other factors.

Getting Started

Web 2.0 is continuing to evolve as a marketing tool, and members of today's construction industry should capitalize on the opportunity. Even if these tools don't seem to fit your style, it is important to recognize younger generations have already integrated Web 2.0 as their means of communicating, learning and sharing. Its applicability will only continue to grow. To further prove the necessity of joining in the conversation, look at the statistics -- they speak for themselves. In April 2004, no one was using online social media. But just five years later, according to compete.com, 91 million were using Facebook, 73 million were using YouTube, 14 million were using Twitter and 12 million were using LinkedIn. And, those numbers are already outdated. According to a study conducted by Cone Business in 2008, 93 percent of Americans believe a company should have a presence in online social media and 56 percent of American consumers feel both a stronger connection with and better served by companies when they can interact with them in a social media environment.

In early 2009, Constructive Communication, Inc. conducted an online social media survey of the AEC industry in which 75 percent of respondents said that they personally interact with online social media regularly, while 80 percent of companies rarely interacted with online social media. Although 76 percent of respondents believe that online social media is a critical medium for their company to get involved in, only 20 percent said they participate in social media outlets representing their company on a weekly basis and 60 percent said they rarely participate from a corporate perspective. The most popular outlet for companies was LinkedIn with 62.5 percent, Facebook with 29.7 percent and blogs with 28.1 percent.

Social Media Policy

Before you dive in and join in the online conversation, create an online social media policy for your company. Here are a few tips:

- The employee should stick to their area of expertise and provide individual perspectives on what is going on at the company
- Post meaningful and respectful comments
- Always pause and think before posting
- Respect proprietary information and content
- Respect confidentiality
- When disagreeing with others, keep it polite and appropriate
- Comply with copyright, fair use, and financial disclosure laws
- When in doubt about posting content, get permission

The policy should also state how employees are expected to engage. Recommend that your employees stay transparent by using their real name and the company name. Other tips for employees would be to write what they know, stay consistent, join in the conversation, add value, create excitement, be a leader and admit a mistake if necessary. Reiterate that employees are responsible for the content they produce. Also, decide if employees are allowed to use this medium while at work and if so, how long while at work.

Another important element of your program is monitoring the conversations. If you are promoting conversations about your company, it is time to start monitoring all conversations. Monitoring services, provided by companies such as Radian6, Cision and Vocus allow you to set up multiple keywords associated with your company.

Online social media is not going away. Even if it is something you choose not to allow employees to participate in, it is crucial that you educate yourself and your leadership team in order to gain optimal positive exposure for your company, as well as alleviate any threats.

Online social media policies are necessary to ensure that you are protecting your organization by setting boundaries for what employees can and should not do online, while also empowering your employees to use social media tools to help grow business and establish your leadership role. Key to success is a policy that guides decisions as well as behavior.

SECTION 4: ALIGNMENT OF SALES AND MARKETING

Sales in the AEC Firm

The Role of Sales in an AEC Firm

It is not enough to simply offer a product or service and hope people buy it. Rather, a leading AEC firm must develop a strategic marketing plan that includes a sales plan and account plans. The sales plan must be a natural extension of the marketing plan. In today's challenging AEC business environment, it's more important than ever to develop a systemic approach to growing business in order to maintain a competitive advantage. Having an integrated marketing and sales strategy in place makes it easier to formulate valuable decisions, create and deliver value and measure performance.

There's no one-size-fits-all sales strategy that will work for every AEC firm. However, there are certain core principles of a successful sales strategy that can be tailored to each business's specific needs. Having a sales approach requires a strategic, systematic method for identifying, satisfying and keeping customers.

Sales...or Marketing?

There are two key components to the business development process: marketing and sales. And any discussion should start with how sales and marketing need to work together. There is often confusion about the difference between marketing and sales. While larger AEC firms may have a sales staff and marketing professionals, too often the roles are combined. Marketing consists of information and knowledge that points to opportunities, and efforts that build awareness and generate leads that result in opportunities. Those efforts and activities include: research, advertising, collateral, direct response, public relations, association involvement, seminars, trade shows and a variety of other activities. The role of marketing is to help condition the environment for a sale by systematically identifying needs and wants, then developing and implementing a plan to communicate potential solutions to satisfying that demand. In contrast, the role of sales is an extension of marketing that is focused on building, developing and maintaining relations that lead to the sale of profitable projects.

Sales efforts in AEC firms include proactive skilled sales professionals who focus on specific activities and relationships that lead to profitable results. These activities include prospecting, presenting capabilities, qualifying opportunities, developing and delivering proposals, negotiating and closing and developing relationships. There are also relationship based sales activities that take place everyday that are executed by engineers, project managers and other in operational roles. Each activity is important and must be aligned to satisfy customer needs

Selling by Market

During the marketing process you determined what markets to focus on how to position. Now, that needs to be incorporated in the Sales Plan. This involves choosing the right strategy, targeting the right customers, targeting specific projects, determining your pricing strategy and matching your sales process to the customer's buying procedure. Implementing a strategic market approach allows your firm to focus efforts and resources on key markets while aligning integrated marketing and sales efforts that get results. Most important, sales and marketing efforts must be consistent with your firm's strategic plan.

As discussed earlier in the chapter on market segmentation, this generally starts with breaking down your clientele into their specific specialties. This step is critical so that specific sales strategies can be developed. Start by segmenting clients into general categories: commercial, industrial, energy, retail, institutional, public and so on. From there, you can break down these broad categories into more specific segments. Next, you can detail the types of structures (big box buildings, banks, educational buildings, high-rises, parking garages, restaurants warehouses, distribution, bridges, etc.) you're likely to encounter in these markets and the audiences you'll be trying to reach with your sales presentation (owners, property managers, engineers, state agencies, etc.). You can further separate markets by grouping them by geographic location, project delivery method, job value or any relevant attributes. Once you've segmented these markets, you can develop targeted sales and marketing approaches that take into consideration the specific needs and challenges of the clientele. Tailoring your approach and messaging will reduce your costs while increasing results, giving you an efficient system for leveraging resources and maximizing sales opportunities.

In the AEC marketplace, the business development process is more challenging due to a variety of factors including: the buying process is complex and takes longer, customers have more information and power and multiple people are typically involved in the buying process. A key element of the sales challenge is to understand the buying process and identify the roles of decision makers and influencers. For each opportunity, different constituents (architects, engineers, owners, and developers, etc.) need to be addressed and they need to be reached with different messages. Identifying each person's role and what is important to them, while challenging, is a critical part of the sales process.

The AEC market is unique in many aspects because while relationships are critical for success, the traditional RFP and purchasing process can quickly commoditize your offerings. A key element of the business development strategy must be differentiation and positioning that leads to competition based on value, not price. This is precisely where the marketing is made personal for the owner.

Sales: Creating Sales Systems, Processes, and Structure

Once your market focus and sales team are in place, you can begin to develop systems that will define how you sell your product or services to clients. These sales activities must be predictable, repeatable, measurable, and designed to facilitate the acquisition, retention or development of account revenue and relationships. Typical sales systems might include standard practices for lead tracking, account management, proposing, determining project capture strategies, estimating the projects and measuring performance. These systems must be designed with the business development professional's perspective, not the operational or design perspective. Further, these systems should enhance the efficiency and effectiveness of the business development staff, not hamper it. In addition, clear lines of communication and interaction need to be established. For example, who is responsible for determining which projects to pursue, estimating, making the Go/No Go decision, developing the proposal and other key business development activities? How can the business development staff engage technical resources in the firm? What role do Project Managers have in maintaining relationships and selling additional work? How can account planning and client mapping lead to customer loyalty? Having regular business development meetings that include discussion of immediate opportunities as well as future sales and marketing activities is critical. These meetings must include representatives from business development, marketing, estimating, construction, operations and engineering. Regular meetings establish a line of communication and allow everyone to be part of the business development process. An important element of sales management is setting expectations, establishing accountably (goals) and setting a process for tracking results (activities and sales).

Resources: Professionals Marketing vs. Marketing Professionals

Business development employees need to be trained and developed, focused and directed, retained for the long term, motivated and rewarded for success. A critical step in developing the firm's business development strategy is determining the type of business development people that are to be representing the firm. However, this is challenging because the majority of people holding marketing or business development roles in the industry do not have formal marketing training; rather, they come from a technical or operations background. After successfully mastering their design or trade skills, these professionals move up the management ranks in their company and often find themselves in a marketing or sales role. Or, they may be viewed as a "people person" by management, and therefore find themselves in a sales role because of their social or networking skills.

There are many benefits to technical staff members holding marketing/sales roles, specifically their ability to intimately understand what they are marketing and selling. However, without sales expertise, their efforts may be haphazard or ineffective. Oftentimes, maybe firms may be too small to support a full-time business development person. In these situations, architects, engineers, and operations people must sell. If this is the case in your firm, the same principles apply.

At any given point, sales involves developing new relationships, maintaining existing relationships, mending strained relationships and re-establishing old relationships. They also function as the vine to the marketplace. They are the key link to understanding changing conditions and expectations. Each firm or company needs to spend time evaluating the marketplace and determining the appropriate sales approach for their market. Then identify the type of skill set required for business development based on the markets being served. For example, an engineering firm with multiple practice areas may have a licensed PE with significant process engineering experience calling on a chemical manufacturing firm and have a non-technical business development person calling on economic development agencies. Ask yourself, do they offer knowledge, credibility, and expertise to bring value to the customers?

Selling What You Know

One of the biggest benefits of having technical people work in marketing and sales in the AEC industry is their ability to clearly articulate a design or construction solution. That is why many firm owners and principles are successful in sales. While the advantages of technical personnel making a presentation or meeting with a client are numerous, absence of marketing and sales skills should not be discounted. Understanding how to properly market, sell and handle client relations are keys to success in today's business environment. Ultimately, the people selling for your firm are charged with filling the funnel with future opportunities. They need to be able to creatively position the company, articulate points of differentiation and sell on value, not price. In today's market, one's ability to bring in business is valued and technical skills are critical to that process but not the only necessity. Buyers are choosing based on personal characteristics and technical skills which are often difficult to separate. In many cases, they are buying a whole team of experts. Although a technical degree offers instant credibility, there needs to be a balance between sales skills, marketing skills and technical expertise. The importance lies in the ability to find the technical answers when needed.

Training is a Requirement

Regardless of whether you are wearing a marketing hat or have a business background, there is one common challenge for marketers in our industry -- lack of training related to marketing and sales in the AEC world. Most business programs, whether undergraduate or advanced, focus on business-to-consumer (B2C) or business-to-business (B2B) with little time spent on professional services. Selling and marketing a professional service is vastly different than selling or marketing a product.

While AEC firms typically invest a significant amount of money in technical training and education programs, they rarely invest in marketing and business development training. In most cases, there is little investment in the soft skills required for business development success such as communication, networking, presentation skills, effective questioning and closing skills. The solution is putting as much effort into training employees on sales and marketing as the latest design or construction trend.

There are a number of excellent business development training programs offered by the Society of Marketing Professional Services (SMPS), Dale Carnegie, Sandler Sales, FMI and a variety of other sources. Business development training should involve specific skills-based training on topics such as communication, building relationships, prospecting, presentation skills, networking, negotiation, closing skills, etc. Training is not a one-time event; rather, it is an on-going process.

It is also important to understand what is meant by B2B. There is a misnomer that B2B means businesses buy from businesses. However, in reality, people in those businesses are making decisions to buy from people in the businesses selling and marketing to them. Simply, people buy from people. Marketing and sales is about positioning your firm and creating opportunities for work.

Building and Managing a Business Development Person or Staff

Whether developed from within or hired from the outside, business development professionals need to be trained and developed, focused and directed, retained for the long term, motivated and rewarded for success.

Most leaders of AEC firms ascended to their positions based on technical or operational accomplishments or capabilities. While they know they need to have a business development person, they are not sure how to lead or manage them. Effectively leading a business development person takes a different approach than managing a project manager, engineer or architect. Leading a business development team is truly a delicate balance between art and science. The process begins with defining focus areas and responsibilities, setting expectations for activities and results (including calls, presentations, opportunities, revenue, profits), tracking and measuring performance and providing feedback and coaching. It is also important to remember that business development people are different than operations and design professionals. However, they need to be part of the team and feel empowered to engage as needed.

Start by taking a look at your existing workforce. Right now, you probably have personnel organized according to operational function; however, creating the most effective sales team will require a slightly different organization of resources.

Even after you have a sales-driven organizational structure in place, your work is not done. Inevitably, you will have to add to your business development workforce at some point and new employees provide an ideal chance to create a more effective sales team. Your strategy should begin before an employee is hired. From job descriptions that identify key competencies to a hiring model that incorporates profiling and testing to accurately assess skills, the recruiting process is your chance to acquire the right person for your company.

Once new employees have signed on with the company, the challenge becomes keeping them there. Reward systems must be tailored for business development professionals. Compensation systems should be designed to reward success and need to include a combination of base salary, commissions and bonuses based on achievement of goals. Providing plenty of opportunities for growth and development—including on-the-job training and mentoring—will not only make employees feel like valued members of the team, but also will benefit your company by creating stronger, more knowledgeable employees.

Sales and Marketing Training for the Technical and Operations Staff

While marketing is the process of attracting clients, sales is the process of having contact and developing relationships with designated customers. In every AEC firm, the technical or operations staff, while not directly responsible for business development, is part of the business development process.

Marketing has been defined as everything seen by the customer. If this is the case, every interaction, every contact the customer has with your firm creates an impression. Everyone in your firm creates an impression anytime they are seen by a client. Each time the client interacts with your firm, they are asking themselves the question: Is this person, or this firm, someone I want to do business with? Every member of your team, from the receptionist to the project manager, has a responsibility to impress the client. An important part of your training process is to help everyone understand their role in the business development process and develop their skills. In many successful AEC firms, the technical and operations staff understands their role in business development. In many cases, they have the closest relationship with the client and are best suited to maintain a pulse with the client. Each person should be trained and held accountable for identifying business development opportunities and bringing in additional work. Through the development of a strategic account management plan and client mapping, firms can identify point of contact with client organizations, assign roles and responsibilities and set goals for maintaining and expanding relationships with clients.

Sales Support

Even the best plan for reaching targeted markets will fall flat unless you have the means to execute it. That's why creating a solid internal structure and maintaining qualified people are also a crucial part of an effective sales strategy. Investments in the sales process must include investments in customer relationship management systems (CRM), such as Sales Logix, Microsoft CRM, ACT!, SalesForce.com, Deltek, or any number of other applications. A dedicated business development representative or team is a significant investment for any firm. These individuals and teams should be focused on high-value activities that lead to results and not be bogged down with administrative activities such as filling out reports, writing letters, assembling proposals and creating PowerPoint presentations.

Can you get sales without a strategy? Of course. But you'll likely find that the work isn't always a great fit for your company or that you're wasting valuable resources and missing opportunities by going after sales leads in an inefficient manner. Don't let acquiring great sales opportunities for your company be a matter of luck—with the right strategy in place, you're guaranteed to succeed every time.

Account Plans and Client Mapping

Just like any other company, architecture, engineering and construction firms need a formal process for managing and developing their accounts. The Account Plan is derived from the Sales Plan which is derived from the Marketing Plan.

Strategic account mapping is a process that is intended to be a flexible, collaborative, and ongoing activity that is both comprehensive and linked to the day-to-day contact and activity management of a firms key and target accounts. The process helps establish management responsibility and accountability for key accounts. For AEC firms, a strategic account program will:

- Provide an understanding of the customer position/situation
- Collect, compile, and analyze information, data, and trends
- Sets specific goals and objectives
- Develop account maps for clients and prospective clients
- Establish responsibilities and accountabilities
- Focus on relationship-building at multiple levels
- Identify opportunities early in the process
- Allocate resources and set measurable objectives
- Help strategically position your firm in the minds of the client or prospective client

The account mapping process will enable business development to better penetrate their accounts, resulting in revenue growth for your firm. The first step involves methodically picking apart your accounts to better understand how they buy. Once the process is started, your firm will be better able to answer a variety of questions about the client. For example, questions should include:

- How do they procure architecture, engineering and construction services?
- What project delivery method do they prefer?

- What is their internal process for capital construction projects?
- What is their internal review and approval process?
- What are budget approval limits and thresholds?
- How do they award projects?
- Do they have existing relationships with other firms?
- Who are the key decision makers and influencers?

The next step is to apply sales strategies that will reveal the account's initiatives, capital and maintenance programs, both funded and unfunded. Then, the account team should look at the various ways that your company can best serve your account and what sales potential these strategies reveal. This will help prioritize where you want to begin and develop tactical sales strategies to employ and align your resources and generate more opportunities and sales. An output should be a visual map of accounts. Ultimately, each team member will see where they fit in the process, including their roles and responsibilities. For example, for client who is an owner of retirement communities, the BD person will have a responsibility to call on the corporate office, while a project manager might have relationships at the facility management level, and an engineer might have relationships at the engineering level. The account map will clearly help everyone see and understand relationships and accountabilities.

Components of the plan should include: a customer mission statement; industry, market, and customer overview; business objectives alignment; position and performance in the account; resource allocation; strategic opportunities and value-based account plan objectives; a plan for using your contact or account management system (CRM); and milestones and metrics.

The Elevator Speech

Although the phrase "elevator speech" has been used in marketing circles for years, it has become a hot topic in the construction industry. While we have all worked hard to get our logo usage consistent, keep our web sites updated and spend the necessary time to create meaningful marketing budgets and plans, the elevator speech still seems to be a bit of a mystery — not to mention quite unnerving. However, by tackling this enigmatic concept and developing the right speech for your company, you'll eliminate the cumbersome and clunky initial sales process.

The phrase "elevator speech" comes from the idea that you should be able to tell a complete stranger everything he or she needs to know about your company in the time it takes to ride an elevator from one floor to another — the length of attention span we typically afford a stranger or new acquaintance. Simply, an elevator speech is a short description — roughly 15 to 30 seconds — of what your company does. This sound bite should succinctly and memorably introduce you.

But beyond simply telling someone what you do, an elevator speech should relate to your audience in a meaningful way. Although it is easy to tell someone you are a concrete subcontractor, for example, it's a bit more cumbersome to share your particular market niche when put on the spot. A good elevator speech should explain the real benefit of the services you provide and how you differ from the competition. For example, are you a concrete contractor who has developed a unique expertise in meeting the needs of owners seeking fast-track solutions? Or have you supplied concrete for more schools than any of your competitors in the last year? By finding the right words, you engage your audience in your business while also proving your expertise.

Developing Your Speech

To develop your elevator speech, you must first identify how your products and services benefit your customers and the marketplace. Your speech shouldn't be a recount of your resume, rather, a summary of what your talents and expertise mean to the person to whom you are talking. Don't feel pressured to tell the whole story as some of the best elevator speeches provide a snapshot of your business but leave the listener begging for more information.

After writing down your differentiator and what it means to others, practice it. Out loud. Often. And then do it again. The most forgotten and ignored step in crafting an elevator speech is rehearsal. Experts have found that one of the strongest signs of performance anxiety is procrastination, and the real reason presenters avoid rehearsals is fear. Contrary to popular opinion, practice will enable your speech to be more spontaneous and flexible, as practice breeds familiarity.

Practice will also help you determine if the words you have selected actually work. The words may look good on paper, but they could sound canned, trite or plain ridiculous when spoken. This is often the case when firms try to memorize their vision statement or tagline and regurgitate the words with little or no connection to the person they are talking to. By contrast, your speech should be something that comes from your heart, makes sense, is easy to remember and rolls off the tongue easily.

How to Gauge Success

Though the elevator speech seems to be one of the most challenging marketing tactics to undertake, it is one of the easiest in terms of gauging success. How do you know if you were successful? It's simple. How did the conversation go after your speech? Did your introduction result in further conversation and questions about your unique selling proposition and industry niche? Do you feel like you engaged your listener and found a way to relate to him or her?

Remember that you never get a second chance to make a first impression. A great elevator speech allows you to position your company as one that has a clear vision and passion for what you do — attributes we all enjoy seeing in others.

Educate and Train Your Entire Staff

Have you ever asked your employees what your company does? Try. Ask a variety of your team members what your company does and you will be surprised by the results. Typically, your team members will have a general idea, but will likely focus on what is closest to them. Ideally, everyone in your firm should understand what your firm does, what type of clients you work for, what differentiates your company and be able to list projects you have completed. This also presents a teaching opportunity. Develop an "elevator speech" template and training process. Once every person understands what you do and how to effectively communicate it, the more effective you will be as a company.

Customer Acquisition vs. Retention and Growth Strategies

There is an important distinction in business development strategies when it comes to acquiring new customers compared to retaining and growing new customers. Each approach requires different understandings, approaches and strategies. Many companies often develop separate marketing plans for customer acquisition and customer retention. Locating customers, attracting them and managing and growing the customer relationships are two very distinct processes.

Many AEC firms enjoy a high rate of repeat business. This is usually the result of good performance and relationships. However, depending upon the shear nature of your business, opportunities for repeat work may not exist. Since the potential for repeat business does exist in many arenas, there is potential for successful retention-focused marketing programs.

Studies show it costs a typical company five times as much to acquire a new customer as it does to retain a current one and most businesses lose 25 percent of their customers annually. By reducing customer base erosion by as little as five percent, a company could add 100 percent to its bottom line. Yet, most organizations allocate six times as much to the process of attracting new customers as they do to holding onto old ones. This is the responsibility of everyone in the firm. Every client interaction leaves an impression of the firm. Positive impressions make it easier to get future work. People like to buy from people they know and trust. Each project must be viewed as an opportunity to impress a client, lay the groundwork for future opportunities and build a relationship. Relationships built on trust and performance are critical to business development success.

As your firm grows, it is essential that you retain and grow as many customers as possible. Marketing programs need to be developed to retain customers and grow the revenue generated from them by offering additional services and products. For example, an engineering firm offering non-destructive testing and engineering analysis services could expand the revenue generated from a client by promoting its design capabilities. Another firm doing design, may capitalize on a clients' desire to have more design/build work by developing a construction management arm. Part of the business development process should involve a formal review of past key customers. For each key customer, a plan to maintain and expand the relationship should be developed.

Section 5: Budgeting, Measurement and ROI

Marketing Budgets

For decades, firms have struggled with the concept of marketing. While you may be trained in the latest technologies, processes and equipment, chances are you don't have any formal education in the latest marketing trends and techniques. This reality has led many firms who have never budgeted for marketing expenditures struggling to track their return on investment and haphazardly respond to marketing opportunities. So, where should a firm who has done little to no marketing for an extended period of time start? Experts state that two to three percent of your total company revenue should be spent on marketing. Properly budgeting for marketing is a large task, but it will provide tremendous dividends and ensure that your company is not left behind as competition continues to increase.

Getting Started

The reason typically cited for not having a marketing plan or budget in place is lack of time to prepare key items. Too often, when tasked with assembling a budget, businesses estimate their overhead, salaries and then gross profit. Anything left over is fair game for technology upgrades, capital expenditures and marketing. However, this method does little to correlate the tactics needed to fulfill your marketing objectives in order to support your sales or branding goals. Setting budgets should focus more on *market opportunities* — identifying the target audiences you need to reach and the anticipated cost to reach them. By determining this number (a number that is realistically tied to your objectives), you can include it in the budgeting process from the beginning and give it the importance it deserves. Further, with this approach, you have the opportunity to develop a baseline to present in defense of your marketing goals. This baseline will identify the mandatory elements of the marketing plan that must be funded, allowing choices to be made on other tactics based on desired results, not dollars.

Begin by identifying what you need to do in order to accomplish your marketing and sales goals in support of your business plan. This involves identifying three to five major goals and outlining the tactics needed to create success. Be sure to identify reasonable, tangible tactics that can be considered in terms of dollars and staffing, but also corporate culture. For example, if improving customer satisfaction is a goal, be sure to outline how you are going to measure that goal beginning with an assessment of the current satisfaction level, as well as steps you will take to improve in this area and a means of measuring satisfaction at the end of the year. Then, conduct a gut-check on all tactics by reviewing whether or not execution is probable. Too often, ideas presented in a marketing plan are solid ideas that have worked for others, yet they don't fit your corporate personality and therefore lack in execution. For example, a marketing plan calling for a personal visit by your CEO to each client with the hopes of improving customer satisfaction is a solid tactic and little argument can be made to the benefit of such a program. However, if history shows the CEO would prefer to stay in the office and leave the customer contact to the project management and sales teams, adjust the tactic to better reflect your corporate personality and culture.

The results of paying extra consideration to developing a marketing plan are obvious. For example, a 25-year-old construction firm had sporadically conducted marketing efforts over their history but had never developed a plan or budget. The firm's president had lost faith in marketing efforts because they ended up costing more than he had anticipated and did not meet the goals of the firm. To combat this, they hired a marketing firm to help them create a clear plan that identified a cost for each tactic. This simple measure allowed them to develop a line item in their yearly budget. They worked their marketing plan for a year while sticking to the budget outline and saw many positive results, including more projects and comments from clients on the effectiveness of their new materials.

Collaborative Budgeting

Probably the most overlooked means of saving money in the marketing budget is the opportunity to collaborate. Although we have all heard this suggestion before, few firms take advantage of the chance to share in marketing expenses. Begin by looking for ways to collaborate with your vendors, associates or other members of your project teams through joint sponsorships, co-op advertising or even shared marketing staff. One example that is easy to implement is project photography. Few firms have the resources (or the desire) to spend top dollar on professional photography for their projects or products. However, if all members of the team were to share in this expense, all would benefit with professional images that can be used for marketing efforts at a reasonable price. This effort also could extend to award submissions, product press releases, event sponsorships, joint presentations at association meetings and more. Be creative in your approach and conduct a brainstorming session with key members of your team to identify other ways to share in expenses.

ROI

One of the most touted terms in the business world is ROI (return on investment). Companies use this term to track a variety of different initiatives but many fail to properly measure the ROI of their marketing efforts. Often, people find it difficult to track these efforts; however, measurement is critical to gaining support from upper management and ensuring your efforts are providing results. A basic understanding of how to measure marketing ROI will help guide efforts and allow for proper planning in future years.

Why Measure?

Branding is another frequently touted buzzword in the corporate world. ROI measurement helps to determine your customers' and prospects' perceptions of your brand image and overall performance.

ROI can help to guide your marketing strategy. By quantifying the achievement of specific goals and targets, ROI can help ensure that your strategy is effective. Many companies who fail to evaluate the effectiveness of specific components of their marketing strategy (advertising, public relations, trade shows) continue to do the same thing year after year, and in doing so inhibit their marketing strategy. For example, consider a construction firm that exhibits every year at a major trade show. This activity accounts for 20 percent of their marketing budget each year, and they continue to go despite the fact that they have received only a few qualified leads from the event. Ongoing evaluation of your marketing tactics will help ensure that your budget is well-allocated. Finally, sales leads generated through individual marketing tactics should be quantified and qualified in order to prioritize marketing investments. This cannot occur if ROI is not being measured.

Now that you know why ROI should be measured, you should also understand the roadblocks to measurement. Often, marketing and sales professionals don't know how to measure marketing ROI. One of the reasons for this is there are few successful examples that can be followed easily. ROI measurement can be even more challenging for construction firms because few examples exist in the market. Internal tracking systems can hinder ROI measurement. Without a marketing plan or unclear objectives, tracking efforts can be very difficult. Finally, firms often do not properly budget for ROI measurement because they incorrectly assume it should be free. While there are many cost-effective ways to track efforts, some will require money and time. Even firms that measure ROI sometimes fail to track every project and only sporadically evaluate marketing efforts.

What can be Measured, can be Managed

Many people are surprised at the number of marketing efforts that can be measured. Marketing can be measured either quantitatively (as a quantity, such as the number of magazine articles written about your company or the number of contacts made by a particular marketing campaign) or qualitatively (in terms of quality, such as the awareness generated by those magazine articles or the influence the output had on a customer's behavior).

For example, measuring placements includes evaluating impressions of how many people received a direct mail piece or saw an advertisement. To quantify this, simply multiply the circulation number by 2.5. To measure ad dollar value for public relations efforts, multiply the length of the article by the ad rates. Since media coverage is earned rather than bought, many experts argue it is more credible than advertising, so it should include an additional multiplier of 10. You can further analyze the quality of public relations efforts by examining the content for prominent placement, picture and key message inclusion and lack of competitors mentioned. Further, each PR piece should be evaluated to see if the majority of information from the press release (including direct quotes, company name, product/service name and contact information) is included.

Measuring awareness requires more effort than a simple analysis of the published piece since it measures outcomes and not just output. Techniques for judging whether customers can recall your marketing communications include surveys, focus groups and other tools. Similar measures can be used to gauge the attitude of customers toward your marketing efforts.

Measuring contacts will require individual evaluations for each type of marketing effort. For example, for PR/advertising/direct mail, you can measure the number of placements or pieces mailed, ad equivalencies, response rate, conversion rate, average order and cost. Similar items can be measured for trade shows, web sites and customer events. Measuring sales is also crucial to assessing ROI. One of the simplest ROI tactics to implement is asking prospects "How did you find out about us?" This enables you to learn from prospects what efforts are working best.

How to Measure Marketing ROI

There are five steps to measuring marketing ROI that you can begin implementing immediately.

1. Set measurable marketing goals that include quantification, deadlines, purpose and an action plan. This goal-setting exercise must take into account historical and forecasted revenues and profits of distinct business segments. Expenditures on various tactical elements should be consistent with these goals.
2. Use multiple measurement tools and methods. For example, tracking impressions, exposures and traffic are important indicators, but ultimately business opportunities, proposals, projects and profits pay the bills.
3. Agree on what is being counted such as identifying what you are trying to measure, what data you are collecting and what data will look like in the final report. When compiling the information and data, use consistent formats. This will also be helpful when tracking and comparing information from multiple years.
4. Use existing measures before creating new ones. The focus should be on efficiency and effectiveness. For example, for search engine marketing and web traffic, Google Analytics offers some very useful and robust tools.
5. Effectively communicate and present your ROI findings to your company. It is important for your team to understand how marketing efforts have a positive impact on the bottom line. Ultimately, company leaders have multiple

and competing choices for investments and expenditures. Demonstrating a return on marketing investments can help ensure future investments. While tracking marketing efforts to show ROI takes some time and money to implement, the results will be realized when you sit down to map out your marketing plan for the next year.

Marketing Performance Dashboards

A marketing dashboard is a tool that enables marketers to measure, monitor, manage and maximize marketing results. Just as a car's dashboard provides the driver with indications of performance, such as speed, temperature, RPMs, fuel level, etc., a marketing dashboard provides marketers with indications of marketing performance. Typically, a dashboard consists of at least three layers:

- At the strategic level, marketing activities are measured against business goals and objectives.
- At the operational level, marketing strategies and systems performance are tracked.
- At the technical level, specific marketing programs, campaigns and initiatives are tracked.

The marketing dashboard helps ensure alignment between marketing performance and business objectives, improves decision making and contributes to the efficiency of marketing investments.

SECTION 6: PUTTING IT ALL TOGETHER

Keeping A Pulse

In every business, there are external factors that significantly impact the business environment. Part of the role of marketing is keeping a pulse on the external environment and competitive activities. This includes financial, regulatory, political, cultural, technological and competitive environments. Changes in any of these environments can have significant implications for businesses. In change, there is opportunity for astute marketers.

With the exponential growth of media outlets, people are bombarded with information. How do you keep a pulse on these environments without being overwhelmed? Start by assessing key factors that are critical to your business success and choose certain things to monitor. One way to do this is by reviewing key general business publications such as the *Wall Street Journal, Business Week* and local business journals. Identify key trade publications focusing on your industry. In construction, start with *Engineering News Record.* Then review publications your customers read. Each industry has a variety of print and digital publications that cover very specific markets. Another great source is the financial community. Financial and investment firms have teams of researchers and analysts that track trends and activity in companies and markets.

Technology has made the research process even easier. Tools such as Google News Alerts allow you to set-up email alerts to inform you when a certain company, product, or even a person is mentioned on the internet. RSS feeds are another way to keep abreast of information. Blog searches such as Google Blogsearch can also be a tool to track trends, but be weary of the source. The emergence of social networking tools such as Twitter and Google Wave allow you to view topics of concern in real-time. Many trade publications offer industry and market reports as well.

Monitoring the Market

In the construction industry, successful firms must continually stay connected to the marketplace to remain competitive. The construction industry is very dynamic. In the last several years, the changes in project delivery methods, sustainability, technology and communication have created opportunities for AEC firms. Part of the marketing function in an organization is to monitor and interpret the trends in the marketplace. Monitoring these opportunities and threats must be an integral part of the marketing process.

Keeping a Tab on Competitors

"Know thy enemy."- Sun Tsu

Every industry has competition. A competitive marketplace creates a healthy business environment. Competitive pressures help drive innovation, differentiation, efficiencies and improvements. A key component of remaining competitive is careful monitoring of the competition. A careful analysis of the competition should focus on competitor's competencies, capabilities and points of differentiation.

Ultimately, the purpose of keeping a competitive pulse is to monitor activities. Through research you will eventually be able to predict what a competitor will do in a certain situation. Developing a deep understanding of competitors and market trends has many benefits. Competitor research efforts should seek to understand:

- Leadership team
- Organizational structure
- Sales and marketing strategies and approach
- Marketing, advertising and pricing strategies
- Competitive strategies and approaches
- Potential responses to your strategies
- How they manage and make decisions
- What they do well and what they don't do so well
- Cost structure
- Compensation structure
- Key customers and relationships
- Goals, objectives and priorities
- Other direct and indirect competitors

- Motivation (publically held, family business, part of a large conglomerate)
- Strategies, objectives, strengths and weaknesses

Where Do You Get the Information?

There are several ways to monitor, track and predict competitive activities and market trends. In addition to the Google News Alerts, monitoring online social media and RSS feeds, media services such as Cision offer information on company news. Dun and Bradstreet, Hoover's, Datamonitor and other services offer profile and financial information for a very low price. Legal and SEC rules require significant disclosure by public companies. Information about private companies may not be readily available but can be obtained with persistence and creativity. Company web sites, press releases and SEC filings provide a tremendous amount of information on the competitor. Competitor's sales people, customers, suppliers, and vendors also are excellent sources of information. Sources include:

- Company web sites
- Search engines, such as Google or Bing
- Social networks, wikis and blogs
- Press releases that offer insight about new strategic initiatives, new products, new services, key hires, new locations, etc.
- Advertisements
- Job postings on sites such as Monster can offer insight into a companies' strategies and initiative
- Customers and suppliers are great sources of information on competitive product and service offering and pricing
- Former employees and current employees
- Vendors
- Trade associations
- Trade shows
- Marketing materials
- Industry paper presentations

How to Get Started

Begin tracking competitive activities and market trends. Build and utilize a tool to store the information and knowledge gathered. Consider both a physical file (brochures, cut sheets, proposals, ad samples, articles, etc.) and electronic sources (files with information collected). Each time you are competing for a project, a significant amount of information can also be gathered from these previous tools during the bidding and proposal process. Involve others in the process as business development, engineering, and project managers can be excellent sources of information. Keep in mind that a key component is ensuring that the information and knowledge collected is current.

Outsourcing Marketing

Depending on the size of your firm, you may not be able to meet all of your marketing needs internally and may want to consider bringing in a third party to handle some or all of the marketing for your company. Working with outside marketing agencies has aided firms in the AEC industry with branding and increasing sales. But outsourcing can be scary, expensive and unproductive if you don't know how to select the right firm, as well as what to expect in terms of fees, contracts and more.

Selecting the Right Firm

The main benefit of working with a marketing or advertising agency is having immediate access to professionals and experts in advertising and marketing communication. However, not all agencies are the same, and therefore it is essential you define exactly what you want in terms of scope and services. For example, most full-service marketing or advertising agencies will work with you on branding efforts; create marketing, sales and advertising materials; as well as handle web site development and electronic marketing. Contrast these services with a graphic design or web design firm, both of which may or may not possess any copywriting expertise. And, while some agencies offer public relations services, another option is partnering with a firm that specializes specifically in publishing, technical writing and the PR process. For smaller projects, you may only need to use a graphic designer or freelance writer to complete a specific part of the project. Knowing what you need upfront is key to finding the right agency — many of those who have unsuccessful histories with outside marketing consultants often admit the firm they hired may have boasted a track-record in one service, such as brochure development but couldn't cut it for the particular service needed.

Further, it is essential that your chosen firm understand the AEC industry. After all, why would you trust your marketing and branding efforts to a firm that knows nothing about your business? Your consultant may have won a variety of awards for television commercials, billboards and point of sales displays in grocery stores. However, if you have to spend hours explaining that your office is made of concrete -— not cement — and that the phrase "latest design trends" does not refer to the fashions on a runway, be careful.

With a strong understanding of the services you desire, as well as the type of firm you need, you should begin your selection process with the same vigor and commitment you would give to hiring a new high-level executive for your firm. For branding, strategic planning or other large marketing efforts, an RFP is completely acceptable when hiring a firm, while interviews may suffice for smaller projects. When interviewing potential firms, meet with the creative staff and senior partners, as well as the account management team. It is critical to develop a level of comfort not only with the agency's creative abilities, but also their ability to link advertising and marketing to business goals and achieve measurable results. Review examples of their past work to see if they fit your style and ask for referrals so you can talk to previous clients about the success of campaigns. Also keep an eye out for red flags: If you see any competitors (or even non-competitors selling to the same industry) on an agency's client list, you should take your business elsewhere. It is not realistic for an agency to serve two competitors honestly and equitably at the same time without compromising someone's success. In the same vein, once you have settled on an agency, it is advisable to ask for an exclusive agreement that prohibits them from working with your competitors. After all, you will need to share your marketing strategy, goals and objectives to arm the firm with the information they need to create success. This information would be catastrophic if leaked to your competitor.

Understanding the Cost of Outsourcing Services

Adhering to the philosophy that owners should not select a firm or product based on low price alone, don't use cost as your only criteria in selecting a marketing or advertising agency. Further, it is important to fully understand how a firm charges for services. Many agencies will propose a retainer, which is ideal if you are working on a variety of projects and want to move forward without the constant back-and-forth with contracts and project estimates. A retainer also will give you more pull in terms of schedule delivery, as the agency will allot time to your account. However, be sure to ask for an accounting of the hours worked each month, as well as goals for the upcoming month, to ensure you are on the same page.

Lump-sum project fees also are common. Expect to pay for a portion of the contract upfront, and be sure to review standard costs that are typically passed onto the client, such as camera-ready artwork, film, couriers, overnight shipping and stock photography and images. Keep in mind if you make a change or modification, the meter is running. A good starting point is a clear definition of the project scope, how external costs will be addressed and include "author's alterations." Also be sure to discuss and finalize who owns the final product. For example, many design firms retain the rights to their graphic concepts and such ownership would permit you from using the artwork for purposes other than original intent or with another consultant.

Ultimately, you want to develop a relationship with a firm that can function as a member of your team. They should understand and add value to your business and be able to offer strategies, counsel, ideas and insight.

The RFP or RFQ

As a member of the AEC industry, your firm is frequently on the receiving end of a RFP. When hiring a marketing firm, the same skill put-forth to respond to an RFP or RFQ should be put into writing the request. Key to success is asking for the right information so you can make an informed decision. Suggested information to gather in the RFP or RFQ to be sent to a firm include:

1. Firm overview
2. Experience in the industry
3. Experience with similar projects
4. Resumes of key persons to work on the account
5. Process and approach for this project
6. Budget and schedule
7. References

Corporate Culture

"The fact is, everyone is in sales. Whatever area you work in, you do have clients and you do need to sell."-Jay Abraham

Most people know that one of the key fundamentals of business growth is delivering what you promise. But if your company is growing, that can be easier said than done. New business often requires additional employees and little time to train them about much beyond the basic policy and software necessities. However, with a documented approach to corporate culture outlining the way things are done at your company, you'll take the first step toward truly integrating new team members.

Why Culture is Important

In the overly competitive construction market, culture is king. Your organization's core values shape every decision you make. Values shape the way you define your strengths, weaknesses, opportunities and threats. They form the backbone of the message you intend to disseminate.

Not only does your culture help guide your everyday actions and delivery of goods and services, but it also helps you differentiate your business from the competition. If a strategic plan is the blueprint for an organization's work and the vision is the artist's rendering of the achievement of that plan, then the culture is the guiding light — the thing keeping you on track. It also tells you when you have derailed. For example, if one of your key messages to the marketplace highlights your commitment to environmental stewardship, make sure your corporate culture embraces such a philosophy. When it comes to culture, your actions will speak louder than words.

For marketers, a key role within the organization is the establishment of a marketing orientation. A marketing orientation can also be considered a customer focus which must be a central tenet of the corporate culture. For the marketing orientation to take root, a consensus must be built at the senior levels of the organization. Once this is established, a proactive effort to market will help ensure the marketing concept is communicated, understood and embraced by employees.

How to Get Started

Documenting a corporate culture begins with a discussion among key stakeholders in the organization about what is important. It is often as simple as asking participants the following: "What do we want to make sure new employees understand about the way we do things and what do we value in terms of our services and our people?" Let the discussion flow as a solid brainstorming session will ensure a variety of ideas are documented. If at all possible, bring in an outside facilitator to help guide the session and keep it on track.

After a healthy discussion has occurred, go back and prioritize the responses. What key themes show up over and over again? What concepts clearly represent the essence of your firm? Next, work your way through the list and figure out how best to capture these concepts in terms of teaching them to the next generation. A workshop for new employees? A letter summarizing the key tenets? Or maybe ongoing training sessions lead by key leaders in the firm. Many have found it is best to create a handbook for new employees that can serve as a reference tool.

For example, one construction management firm was troubled with new employees, all of whom were technically competent engineers but were missing key details when they reviewed drawings simply because they didn't understand "the way things are done in the firm." Since this firm is known as being extremely detail-oriented — a valued part of their culture by the firm's leadership and clients alike — they recognized they had to find a way to teach their culture to their growing team. As such, they developed an outline of tasks to guide the review efforts and provide discipline to tasks such as building a project schedule. A series of questions ensured that all new employees know how to look at a project in front of them.

Another key example is developing a standard for CAD drawings or other deliverables. While all of your drawings may look alike right now because your two drafters have worked together seamlessly for a decade, throw two new hires and a plethora of new projects into the mix and you'll likely have a new format spitting out of the plotter by week's end. An established, documented process for completing drawings will ensure that they always remain consistent, no matter how many new variables are introduced. Remember, each contact with the customer is an impression of your company.

Documenting Success

Passing down the corporate culture goes well beyond documenting the approach a firm takes to project development and delivery. In today's harried business world, it is essential to share corporate values with new members of your team. For example, if recycling is an important part of your culture or partnering with community organizations is valued, make sure everyone knows it. Better yet, make sure they know why you value such activities and how it fits in with your cultural mindset. The clearer you are on defining the culture and explaining the reasoning behind the belief system, the more likely new employees will embrace the same doctrine. Such buy-in not only ensures a solid corporate culture, but is likely to result in a profitable business as well.

Advisory Boards

One of the key elements of business success is the ability to change. However, it's often hard to figure out why, what and how to change without any outside ideas. That's why many companies in the design and construction industry have formed advisory boards. Whether formal or informal, an advisory board can help gain a fresh perspective and eliminate making the same mistakes others have made in the past.

Unlike corporate boards, advisory boards have no fiduciary responsibility and their advice is non-binding. Separate from a shareholder or investor board, an advisory board can provide you with direction that isn't necessarily tied to the bottom-line or a stake in the outcome. Rather, an advisory board can provide outside guidance on key issues affecting your business — marketing, sales, staffing, succession planning and more.

Jim Joyce, president of H.R. Gray — a management and consulting firm serving the construction industry exclusively for public projects — also has seen the benefit of retaining key advisors.

"I formed an advisory board because we didn't have any in-house experience in operating a company although we are very experienced in the work we obtain," said Joyce. "I felt it was important to learn from others and wanted to structure our financial reporting so it looked like what other businesses were doing and what the financial marketplace was used to."

To assemble his board, Joyce contacted his local business publication and relied upon their database of people interested in serving on advisory boards. He reviewed and selected candidates based on their personal experience and familiarity with the industry. Although it took eight months to assemble the right players, they now have a chair with strong CEO skills, a CPA with a strong construction background and a political consultant/lobbyist with a strong understanding of public works contracting practices.

"It helps to have a different perspective," he said, "I think it provides a more complete review of how we run our company. Subcontractors, suppliers and clients can give valuable insight because they see many firms in the same position as yours and can give you a best-practice viewpoint."

While some advisory boards are hands-on, meeting once a month or more, many advisory boards come together only a few times a year. Still other companies operate with advisory boards that never meet; rather, the president of the company has a list of go-to people for testing ideas. Such is the case for Hardwire, LLC.

Founded in 2002, Hardwire is a privately held company that produces high-tensile steel reinforcements for the broad composites industries and manufactures armor to protect critical domestic facilities and infrastructure. Hardwire products, often combined with concrete, can be found in blast-resistant structures, reinforced and corrosion-resistant flooring overlays, concrete repair retrofits, reinforced piping and many other applications. George C. Tunis III, chairman & CEO of Hardwire, has an extensive network of advisors whom he calls upon for specific needs and strongly encourages members of the industry to interact more as it benefits everyone.

"I think the key to really high-caliber advisors is to never waste their time with meetings that may cross too many lines of expertise," said Tunis. "Instead, focus your advisors' time by always keeping them up to speed with frequent updates or calls, call on them when their expertise applies and the subject fits their passion or business interests. The best advice comes when the subject benefits both parties, even when there is not an immediate business fit. Then be sure to say thank you in the most personal and appropriate way."

The difference between a formal advisory board and constant, continuous networking with a circuit of trusted, capable individuals is simply format, noted Tunis.

"Some people are more comfortable with the structure of an advisory board, but I, like most entrepreneurs, am just too impatient to wait for a regularly scheduled meeting," said Tunis. "And more important, when it comes to protecting soldiers overseas with Hardwire's ballistic- and frag-resistant panels, or developing lighter and more effective vehicle armor, time is critical. Our outside advisors bring real-world experience to the solutions we develop and for that we are beyond grateful."

Assembling Your Board

The key to success in selecting your board is assembling a group of business owners or executives whom you respect, as you are more likely to listen to their ideas. Also do your best to pick companies with complimentary services or products. Not only does such a grouping provide a chance for interaction and leads among the members of your board, but a group of competitors will likely be tight-lipped.

Although advisory boards are typically unpaid positions, compensating the members in some fashion is appreciated — whether through in-kind service, gift certificates or even leads. Also, be cognizant that while their greatest gift may appear to be their ideas, the real gift is their time. As such, it is crucial that you establish an agenda, provide ample background information and run a structured meeting. The same goes for an informal group of people you call upon: Even if you are running an idea by them on the phone, be fair to their time by scheduling the conversation in advance and giving them a chance to think about the topic before you talk.

Everyone's a Marketer

"Marketing is far too important to be left only to the marketing department."

--David Packard, co-founder of HP

It is often said marketing and sales are roles held by all members of the firm. At its core, this is correct. Simply, marketing is too important to be left to the marketing department. Even if you wanted to relegate marketing to one department, it isn't possible as anyone who interacts with customers leaves an impression about the company – from the receptionist who answers the phone to the lowest level field worker. What employees wear, say and act all serve as the outward marketing arm of your firm.

Recognizing this to be true, it is important to help everyone understand their role in consistently branding the firm. Marketing needs to be a mindset. Too often, marketing personnel are just viewed as "overhead," and while some may appreciate their role, marketing is sometimes viewed as a necessary evil. Key to sustainable growth for an organization is helping non-marketers see the importance of the marketing function.

It begins with communication about your marketing goals, strategies and activities. There is nothing worse than a technical staff member being caught off guard when a client mentions an ad, article, brochure or sponsorship activity that he/she hasn't seen. Such items should be circulated so all team members not only have the knowledge of such activities, but can help promote them as well. But, communication is two-way. It is important for marketers to regularly meet with technical staff members to ask their opinion about industry trends, marketing campaigns and key strengths of the firm so as to include that information in future marketing pieces.

In addition to communication, education must occur. As noted early in this manual, most technical staff do not have any marketing or sales training. Many firms have integrated lunch-time brown-bag sessions into their educational efforts to teach technical staff about presentation and public speaking skills, dealing with different personality types, how to network and more. Such efforts educate technical staff members in a non-threatening manner.

Finally, it is important to develop processes and tools that make the most of the technical staff's time, so marketing isn't viewed as a burden. The discontent between marketing and billable staff will likely occur if marketing is too demanding of time or are disorganized in their requests. Key to success is involving the team early in the process and showing respect for their time with a thorough process for gathering input in a timely manner.

Marketing in a Down Economy

"When written in Chinese, the word 'crisis' is composed of two characters. One represents danger and the other represents opportunity." -John F. Kennedy

Like every other industry, the AEC industry has its ups and downs. Typically, the design side is the first to feel the effects of difficult economic times and the first to come out of it. However, while many tighten their belts and cut their expenditures, in challenging economic times and competitive conditions, marketing is more important than ever. While dollars may be tight, studies show that a consistent presence in the marketplace now will pay dividends in the future. Challenging economies presents excellent opportunities to review your marketing efforts and ensure you are acquiring new and retaining existing customers. Further, many have shown that marketing dollars invested in a down economy are actually more valuable than those spent in a good economy as you gain greater market presence and are seen as being a strong, solid company.

Regardless of the economy, marketing strategies must be constantly reviewed to ensure business goals are being met. This is just as true in a down economy as it is a great time to take advantage of limited marketing spending by others, which improves your chances of getting noticed and standing out. In other words, turn the challenging economy into an opportunity. Although it may seem prudent or even appear to be a good business decision to cut back on marketing costs to "save money," it is very important to maintain consistency and stay true to your marketing goals. After all, the need for building a brand, creating awareness and generating leads for opportunities doesn't go away when the economy is sluggish. In fact, it is more important than ever in a down economy.

When your competitors choose to cut back on their advertising and marketing budgets, such a scenario results in less of a presence from your competition in the marketplace and a greater impact will be realized with your efforts. Further, with decreased revenue, many publications may be willing to negotiate lower prices or value-added package deals, editorial opportunities or merchandising opportunities.

It also is a great time to invest in public relations as publications may tighten their freelance budgets and are likely looking for article content. Another opportunity may be creative partnerships with vendors such as trade show hosts, printers, advertising agencies and more as these firms want to keep their own employees busy during down times. Digital media is another opportunity. Be sure to take advantage of the efficiencies of digital which include lower costs and improved tracking opportunities.

Marketing and advertising during a sluggish economy also boosts the confidence that customers and potential customers have in your company as your consistent market presence will deliver the message that you are a strong, viable, stable company. Such a message contributes to long-term brand building and supports your differentiation strategy. And, when the economy builds again, the top-of-mind awareness you've built will result in a sustainable competitive advantage.

In sales, there typically is not a better source of future revenue then your current customers. And, this holds true in tight economic times. Your current customers already have a relationship with you and hopefully you have left a positive association with your brand for them. Continue to invest in integrated marketing programs to maintain constant contact with your customers. Find more ways to add value by offering current customers a new or expanded offering of services and products. Offering complimentary services and products is an excellent way of providing more value to your customers when they are struggling. Again, the goal is to find opportunities to provide more value and forge deeper relationships.

While downturns in the economy are challenging, they also force us to look at activities and initiatives and make tough decisions. In reality, this is a very healthy process and forces companies to look at the return on marketing investment. A critical evaluation of marketing investments will result in greater efficiency and more effective programs.

Consistency is essential in marketing as consistent investments in marketing yield consistent results in terms of lead, opportunities, closed sales and profits. Resist the urge to slash marketing investment and seize the opportunity build your brand, generate more opportunities and drive more profitable revenue.

Specific strategies in difficult economies:

When there is considerable uncertainty, there are several approaches and strategies AEC firms can take to weather the economic situation and be positioned for opportunities when the economy improves.

Focus on Key Customers: Continue to maintain and develop customer relationships. Focus business development efforts on key clients and building loyalty. Also protect your key clients from competitors. Communicate regularly with your clients and convey a message of strength and stability. Customers want to work with a vendor they can trust and rely on in difficult times. Also consider handling different projects for key customers.

Communication: Communicate regularly and clearly with your employees. Your employees are barraged with negative media messages about the economy so you have an important role to communicate with your employees candidly about your business and marketplace. Consider engaging them in finding cost-saving strategies.

Change the BD Strategy: Many companies have cancelled or delayed large capital and maintenance projects. Demand is shrinking. While your customers or prospective customers have financial constraints, this does not mean all projects will be cancelled. Your job is to understand their financial condition, verify funding sources and understand how you can tap their budgets. Consider breaking a large project into smaller projects, offering some incentives or bundling work. Be flexible by working with your customer to create win-win solutions. Also seek to build diversity and consider other types of projects your firm may not have considered in the past.

Modify Your Marketing Messaging: A financial crisis can turn the business world upside down. If your company is fortunate to have a long history, focus on your stability in your marketing messages.

Focus on Effective Marketing: Many companies decrease marketing investments during tough economic times. In fact, marketing is one of the first things to get cut. This is a great time to invest in marketing. Publishers are typically willing to negotiate favorable terms. Fewer people are advertising so your ads and messages have a better chance to cut through the clutter. Also consider shifting dollars to web marketing initiatives such as SEO and Web 2.0 efforts.

Focus on Niches: Identify key niches or market segments where you have a strategic or competitive advantage. The reality of a challenging economic situation is that there are more bidders on projects and prices typically go lower. Offering value-added services and having a clear focus on markets that are not solely price driven will help you weather the storm.

Upgrade Your Talent: During challenging economic times, talent supply exceeds demand in the construction industry. Talent, including engineers, project managers and business development people may be available. Take the opportunity to add resources that will strengthen your firm.

References

1 Kotler, P. 1999, *Kotler on Marketing: How to Create, Win and Dominate Markets*, Simon and Schuster, London, UK.

2 Porter, M. 1985, *Competitive Advantage: Creating and Sustaining Superior Performance*. Free Press, New York, NY. P. 12.

3 Reis, A. & Trout, J. 2001. *Positioning: The Battle for Your Mind.* McGraw Hill, New York, NY.p. 19.

4 Kotler, P.2001, *Marketing Insights From A to Z*, John Wiley & Sons, Hoboken, NJ. P. 108.

5 Reis, A. & Trout, J. 2001. *Positioning: The Battle for Your Mind.* McGraw Hill, New York, NY.P. 24.

6 Imber, J., & Toffler, B., *Barron's Business Guides: Directory of Marketing Terms.*

7 Kotler, P.2001, *Marketing Insights From A to Z*, John Wiley & Sons, Hoboken, NJ. P. 4.

8 Reis, A. & Reis, L. 2002. *22 Immutable Laws of Branding.* Harper Collins, New York, NY.

9 Godin, S., 1999, *Permission Marketing*, Simon and Schuster, New York, NY.

APPENDIX

Marketing Plan Basics

There are numerous resources available that provide detailed information on the marketing plan structure. In addition, there are a variety of formats that can help you reach your objectives. Here, we've provided a basic structure for your marketing plan.

Executive Summary

High level overview of your marketing plan and expected outcome

Situational Review & Analysis

SWOT Analysis

Company Analysis

Market Research and Analysis

Customer Analysis

Competitor Analysis

Goals/Objectives

Linked to the overall business plan

Specific and measurable

Opportunities should fit your strengths

Goals/Objectives need to be rank and prioritized

Must have specific time frames and measurements

Strategies

Specific actions of how you are going to reach the Goals/Objectives

Differentiation, position and messaging

There are different paths to achieving each objective

Segmentation and strategies by market

Strategy is about making a strategic choice and following a course of action

Tactics

Details of the strategy

The Marketing Mix

Implementation of the "4 Ps"

Communication plan

Specific marketing initiatives (advertising, direct response, public relations, seminars, web, collateral, etc.)

Outline specific strategies by market

Marketing activity calendar

Linkage to the sales and account plans

Budget

Cost for each planned activity

Specific mechanisms for tracking costs

Controls

Tools & metrics to measure success

ROI tools

Appendix

Exhibits

Schedules

Financial Forecasts

SWOT Analysis Basics

The SWOT analysis is a great tool for analyzing a situation and helping with the decision-making process related to marketing. This process helps firms align with their markets. The term SWOT stands for Strengths, Weaknesses, Opportunities, and Threats. Strengths and Weaknesses are internal to the organization, and Opportunities and Threats are external to the organization. The SWOT helps develop a framework for building a marketing strategy. The SWOT process can be done for the entire company, a particular business line, or a market. SWOT analysis can also be particularly useful in evaluating competitors.

SWOT analyses are most effective when built by a small group of people that represent different parts of the organization. This helps bring perspectives and viewpoints that can be valuable.

Below is a SWOT with example topics. These will very by business.

Strengths (Internal)	**Weaknesses (Internal)**
Capabilities Brand Resources Reputation Competitive Advantage Staff Experience Processes Systems Culture Management Market and Sales Approach	Financials Funding for Marketing Vendors Gaps in Capabilities Gaps in Experience Management Holes Leadership Voids Sales capability
Opportunities (External)	**Threats (External)**
Industry Trends Client Trends Emerging Markets Vertical & Horizontal Markets Niches Geographies Competitor Vulnerabilities	Competitor Activities Competitive Positioning Political Environment Economy Market Demand New Technologies

After identifying the strengths, weaknesses, opportunities and threats, rank the top items in each category in terms of how they will impact your business. This process will help identify strategic priorities. Once the SWOT analysis is complete, it can be used for formulation of strategies. To start this step, you will need to ask a few key questions about the top items you identified.

- How can we leverage and capitalize on each Strength?
- How can we improve upon each Weakness?
- How can we capitalize on and benefit from each Opportunity?
- How can we minimize or eliminate each Threat?

This information gathered in this process will be very useful in developing your marketing plan.

Communications Plan Basics

Positioning Statement

Without direction or focus, a business or organization runs the risk of delivering too many messages to its target audiences. Often times, these messages even conflict. Without a clear position statement, the messages don't mean much or say anything. From a management perspective, *positioning* is the heartbeat of an effective communications plan. A well-crafted positioning statement defines your company's direction and answers seven essential questions:

1. Who you are?
2. What business you're in?
3. For whom (what people do you serve)?
4. What's needed by the market you serve?
5. Against whom do you compete?
6. What's different about your business?
7. What unique benefit is derived from your product or services?

Don't confuse a positioning statement with market position. While market position relates to how you are perceived in the minds of your prospects, a positioning statement expresses how you wish to be perceived. It is the core message you want to deliver in every medium.

Mission Statement

A mission statement is an expression of a company's history, managerial preferences, environmental concerns, available resources and distinctive competencies to serve selected target audiences. It is used to guide the organization's decision-making and strategic planning.

Vision for the future

Many confuse the mission and vision statements. A vision is a guiding image of success formed in terms of a contribution to society. If a strategic plan is the "blueprint" for an organization's work, then the vision is the "artist's rendering" of the achievement of that plan. It is a description that conjures up a similar picture for each member of the group of the destination.

Audience

The audience is the number and/or characteristics of the persons who are exposed to a particular type of advertising media or media vehicle. Who does your organization come into contact with? Who does your organization serve or intend to influence? When considering your audience, keep in mind the destination you want them to reach following exposure to your message. What do you want your audience(s) to: know, believe, feel and do as a response to your message?

There are three different kinds of audiences to consider for communications efforts:

- Primary Audience – Also known as the target audience; it is a subset of the general population that you are trying to reach. Who are you trying to reach with your message? Is it internal or external? Describe your primary audience as specifically as possible and include age, gender, education, etc. It is also important to consider the media your audience reads and is exposed to. Having a clear picture of your intended audience in mind will assist you in making good decisions about how to reach your audience. When you want your message to reach different audiences, prioritize the importance of each audience. The most important audience is your primary audience.
- Secondary Audience – Sometimes a message is intended for more than one audience, or subset of the population. The secondary audience also includes the people who can influence your primary audience.
- Tertiary Audience – Another audience, third in place, order, degree or rank.

Role of the Communications Plan

The role of the Communications Plan is to guide the efforts of the organization with regards to communicating with identified audiences. Your Communications Plan is your roadmap to success using your intended message(s). A Communications Plan helps you deliver messages that are clear, memorable and effective. With this written plan to guide your communications efforts, you will be able to identify: what you want to accomplish; how to reach the communications goals you have set; the tools you need to achieve success; and how to measure the effectiveness of your communications efforts.

Situational Analysis

A situational analysis is commonly assembled by reviewing internal (Strengths and Weaknesses) and external factors (Opportunities and Threats) through a process called a Strengths, Weaknesses, Opportunities and Threats (SWOT). A SWOT analysis (see page 183) is a simple framework for generating strategic alternatives from a situational analysis. Strengths can serve as the foundation for building a competitive advantage and weaknesses may hinder it. By understanding these four aspects of your current situation, you can better leverage your strengths to correct your weaknesses, capitalize on opportunities and avoid threats.

Assumptions

Assumptions are the ideals you believe to be true, without needing or having proof, about your organization.

- External forces detail the external forces that must be taken into consideration with regards to the communications efforts. These are the forces outside your organization that you can't control, such as the economy. In your situational analysis, opportunities and threats are external forces on your organization.
- Internal limitations detail the internal factors that must be considered, such as lack of manpower, dwindling budgets, etc. These are the forces inside your organization that you can control. In your situational analysis, weaknesses are internal limitations.

Key Messages

Key messages are the strategic points you want to deliver with your communication efforts. It is wise to have only three or four, as readers won't remember more than that.

Key messages are the core of your writing - they open the door to direct communication with your audience and bridge what your audience already knows and where you are trying to take them.

You have a point to make—whether to educate, discuss, promote or advocate. Within every text, key messages are the points you want your audience to remember and react to. They are *The message* or the essence. Within all your writing, key messages keep your marketing and sales efforts on-track with what you are trying to accomplish. Readers should always come back to your key messages. More important, your key messages are your call to action – these messages tell your audience what you want them to do. Key messages are a means to an end. They assert your viewpoint. Key messages are opinions that you can back up with proof and case examples, which you demonstrate within your marketing efforts. Key Messages get your audience curious about what you have to say. Curiosity is the first step to participation.

Internal Values and Their Impact

Culture is key. In addition to spelling out key messages, the culture of your organization is accounted for in an internal value statements. For example, if your key messages highlight your commitment to environmental stewardship, but this is not something that is part of your culture, your actions will speak louder than words.

Your organization's core values shape every decision you make. Values shape the way you define your strengths, weaknesses, opportunities and threats. They form the backbone of the message you intend to disseminate. An internal assessment of your organization's core values prior to establishing your Communications Plan will ensure congruence in your message and your organization's actions.

Communications Goals

What is the desired response to our activities? A successful communications plan will have realistic, clear and action-oriented goals. Goals should be specific and measurable in order to facilitate an accurate evaluation and ROI measurement.

Strategies

A strategy is the plan of action for reaching your communications goals. Strategies answer the question, "What will we do to achieve our communications goals?" When establishing a strategy, consider the communication tools at your disposal and determine which ones are best suited to achieving your goals. Usually, a variety of communication methods will deliver the best results.

Tactics

While strategy is the plan of action intended to accomplish a specific goal, tactics are the means to fulfill the strategy. Tactics should detail the action, timing, resources needed, persons responsible, budget, etc. Tactics answer the question, "How will we achieve our communications goals?" Tactics employ every means to achieve the intended result. In a list of tactics, each tactic employed should also include success criteria for that tactic.

Evaluation

Most plans fail to include an evaluation or means to measure ROI. Were we successful? The key to successful evaluation is setting measurable goals for your communications plan. Your evaluation may include:

- A monthly report on work in progress
- Formalized reports, presented at meetings
- Periodic briefings of the CEO and other department heads
- A year-end summary for the annual report

An evaluation should include recommendations for the future. Based on the effectiveness of the current communications plan, what are the upcoming goals for your organization?

Printed in Great Britain
by Amazon.co.uk, Ltd.,
Marston Gate.